The Beginner's Guide to Raising Healthy, Happy Chickens

Learn the Basics of Chicken Care, from Egg to Coop

Danielle Schultz

Copyright © 2024 by Danielle Schultz

All rights reserved. No part of this publication may be reproduced, stored or transmitted in any form or by any means, electronic, mechanical, photocopying, recording, scanning, or otherwise without written permission from the publisher. It is illegal to copy this book, post it to a website, or distribute it by any other means without permission.

Introduction..1
  Welcome to the World of Chicken Keeping.............. 1
  Why Raise Chickens?...4
  What You'll Learn in This Book................................ 5
Chapter 1: Getting Started with Chicken Keeping.....8
  Choosing the Right Breed..8
  Understanding Basic Chicken Behavior.................. 11
  Setting Up Your Chicken Coop............................... 13
Chapter 2: Acquiring Your Chickens....................... 16
  Buying Day-Old Chicks vs. Adult Birds................... 16
  Selecting Healthy Chickens..................................... 19
  Transporting Your Chickens Safely........................22
Chapter 3: Chicken Nutrition and Feeding...............25
  Understanding Chicken Nutritional Needs..............25
  Choosing the Right Feed.. 28
  Supplementing with Treats and Scrap.................. 30
Chapter 4: Health and Wellness.............................. 34
  Preventative Care: Keeping Your Chickens Healthy... 34
  Identifying Common Health Issues........................ 37
  Administering First Aid to Sick or Injured Chickens 40
Chapter 5: Egg Production and Care........................43
  Understanding the Egg-Laying Process................. 43
  Collecting and Handling Egg................................... 45
  Maintaining Egg Quality...48
Chapter 6: Managing Your Flock...............................51
  Handling and Socializing Your Chickens................ 51

Dealing with Behavioral Issues............................54
Integrating New Birds into the Flock....................57
Chapter 7: Coop Maintenance and Cleaning........... 61
Cleaning and Sanitizing the Coop....................... 61
Preventing Pest Infestations................................64
Routine Maintenance Tasks................................66
Chapter 8: Seasonal Considerations....................... 70
Winter Care: Keeping Your Chickens Warm and Healthy...70
Summer Safety: Protecting Chickens from Heat Stress...74
Fall and Spring Preparations.............................. 76
Chapter 9: Breeding and Incubation........................79
Understanding Chicken Reproduction.................... 79
Incubation Process.. 80
Incubating Eggs.. 82
Caring for Chicks.. 84
Chapter 10: Free-Range and Pasture-Raised Chickens..87
Benefits of Free-Range and Pasture-Raised Chickens... 87
Considerations for Free-Range Management........ 90
Creating a Safe Outdoor Environment...................93
Chapter 11: Sustainable Chicken-Keeping Practices.. 96
Using Chicken Manure for Fertilizer...................... 96
Integrating Chickens into Permaculture Systems... 98
Benefits of Integrating Chickens into Permaculture Systems...99

Key Strategies for Integrating Chickens into Permaculture Systems..100
Complementary Farming Practices...................... 100
Chapter 12: Legal and Regulatory Considerations..... 103
Zoning and Local Ordinances............................... 103
Compliance with Health and Safety Standards.... 105
Permits and Licenses.. 108
Chapter 13: Troubleshooting Common Issues....... 110
Solving Egg-Laying Problems............................... 110
Dealing with Predators.. 116
Chapter 14: Chicken Products and By-Products... 119
Utilizing Chicken Manure...................................... 119
Harvesting Feathers and Down............................ 123
Processing and Utilizing Chicken Meat................ 126
Conclusion: Reflecting on Your Chicken-Keeping Journey..131
Further Resources and Next Steps...................... 133

Introduction

Welcome to the World of Chicken Keeping

Raising chickens is an enriching endeavor that offers a deep connection to nature and a sustainable lifestyle. As you embark on this journey into the world of chicken keeping, you are stepping into a realm filled with wonder, learning, and fulfillment. Whether you're a seasoned farmer or a novice enthusiast, the experience of caring for these feathered creatures is bound to captivate and inspire you.

The allure of chicken keeping transcends mere practicality; it encompasses a holistic approach to sustainable living. Beyond the provision of fresh eggs and meat, chickens contribute to soil health through their natural fertilization process, aid in pest control, and even provide companionship with their quirky personalities. In essence, keeping chickens is not just a hobby; it's a way of life—a harmonious integration of humans and animals within the fabric of the natural world.

Amidst the hustle and bustle of modern life, tending to a flock of chickens offers a respite—a return to simplicity and self-sufficiency. In the tranquil ambiance of a rural homestead or the cozy confines of an urban backyard, the rhythmic clucking of hens and the contented scratching of their feet against the earth evoke a sense of serenity and connection to the land. Indeed, in the fast-paced digital age, the act of caring for chickens allows us to reconnect with the timeless rhythms of nature.

As you embark on your journey into the world of chicken keeping, it's essential to approach this endeavor with enthusiasm, curiosity, and a willingness to learn. Whether you're drawn to the prospect of farm-fresh eggs, the joy of watching fluffy chicks hatch, or simply the companionship of feathered friends, raising chickens offers a multitude of rewards and experiences.

One of the most appealing aspects of chicken keeping is its accessibility. Unlike larger livestock such as cows or horses, chickens require minimal space and resources, making them ideal for both rural and urban settings. Whether you have acres of land or a small backyard, there's a chicken-keeping setup to suit your needs and circumstances.

Beyond the practical benefits, however, lies a deeper connection to the natural world. In caring for chickens, we are reminded of our role as stewards of the earth—responsible for the well-being of all living creatures. From providing nutritious feed and clean water to ensuring comfortable living conditions, each aspect of chicken keeping reflects our commitment to nurturing and protecting the animals under our care.

Moreover, the bond between humans and chickens goes beyond mere utility; it is one of mutual respect and companionship. Chickens, with their endearing personalities and social dynamics, have a remarkable ability to forge emotional connections with their human caretakers. Whether they're following us around the yard, eagerly awaiting treats, or simply basking in the sun, chickens have a way of brightening our days and reminding us of the simple joys of life.

The world of chicken keeping is a tapestry of experiences—rich, diverse, and endlessly rewarding. From the first tentative steps of acquiring your flock to the daily rhythms of feeding, cleaning, and tending to their needs, each moment spent in the company of chickens is an opportunity for growth, learning, and connection. So, as you embark on this journey, embrace the challenges and delights that lie ahead, and welcome the world of chicken keeping with open arms.

Why Raise Chickens?

The decision to raise chickens is rooted in a myriad of practical, emotional, and environmental considerations. At its core, raising chickens offers a sustainable and self-sufficient approach to food production, providing a source of fresh eggs, meat, and fertilizer for individuals and families. In an era marked by concerns about food security and environmental sustainability, the act of raising chickens empowers individuals to take control of their food supply and reduce their ecological footprint.

Beyond its practical benefits, raising chickens also fosters a deeper connection to the natural world. In an increasingly urbanized society, the act of caring for chickens offers a tangible link to the rhythms of rural life and the cycles of nature. From the gentle clucking of hens in the morning to the sight of fluffy chicks exploring their surroundings, raising chickens provides a daily reminder of the beauty and wonder of the natural world.

Moreover, chickens possess a remarkable ability to forge emotional connections with their human caretakers. With their endearing personalities and quirky behaviors, chickens have a way of capturing our hearts and enriching our lives in unexpected ways. Whether they're eagerly greeting us at the coop door or contentedly pecking at the ground, chickens have a unique charm that brings joy and laughter to those who care for them.

From a practical standpoint, raising chickens offers numerous economic benefits as well. Compared to store-bought eggs and poultry, home-raised chickens are often more cost-effective, allowing individuals to save money on their grocery bills while enjoying higher-quality, fresher products. Additionally, the by-products of chicken keeping, such as manure and feathers, can be utilized in garden composting and crafting projects, further enhancing the sustainability of the endeavor.

In essence, the decision to raise chickens is a multifaceted one, influenced by considerations of practicality, sustainability, and emotional fulfillment. Whether you're motivated by a desire for fresh, wholesome food, a connection to nature, or simply the joy of caring for living creatures, raising chickens offers a wealth of benefits and experiences that are bound to enrich your life in countless ways.

What You'll Learn in This Book

As you embark on your journey into the world of chicken keeping, this book serves as your comprehensive guide, providing you with the knowledge, skills, and confidence to successfully raise healthy, happy chickens. From the basics of chicken care to advanced techniques for managing your flock, each chapter is designed to equip you with the information you need to thrive in your chicken-keeping endeavors.

In the opening chapters, you'll learn about the fundamentals of chicken keeping, including how to choose the right breed for your needs, set up a safe and comfortable coop, and acquire your first flock of chickens. With practical tips and step-by-step instructions, you'll gain the knowledge and confidence to get started on the right foot and ensure the well-being of your feathered friends from day one.

As you delve deeper into the book, you'll explore topics such as chicken nutrition and feeding, health and wellness, and egg production and care. You'll learn how to provide your chickens with a balanced diet, identify and treat common health issues, and optimize egg production for maximum freshness and quality. With expert advice from experienced chicken keepers, you'll be well-equipped to address any challenges that may arise and keep your flock happy and healthy year-round.

Furthermore, this book covers essential topics such as managing your flock, coop maintenance and cleaning, and seasonal considerations. Whether you're dealing with behavioral issues, preventing pest infestations, or preparing your chickens for the changing seasons, you'll find practical guidance and troubleshooting tips to help you navigate any situation with confidence and ease.

In addition to practical advice, this book also explores the broader implications of chicken keeping, including its environmental impact, legal and regulatory considerations, and potential for sustainable practices. You'll gain a deeper understanding of the role of chickens in sustainable agriculture and how you can contribute to a healthier, more resilient food system through your chicken-keeping efforts.

Chapter 1: Getting Started with Chicken Keeping

Getting started with chicken keeping is an exciting and rewarding endeavor that opens the door to a world of fresh eggs, lively clucks, and the soothing rhythms of rural life. Whether you're a seasoned farmer looking to expand your flock or a novice enthusiast taking your first steps into poultry-raising, this chapter serves as your comprehensive guide to navigating the ins and outs of chicken keeping with confidence and ease. From selecting the right breed for your needs to setting up a safe and comfortable coop, you'll find all the information you need to embark on your chicken-keeping journey with success.

Choosing the Right Breed

When it comes to choosing the right breed of chickens for your flock, there are a multitude of factors to consider, from egg production and temperament to climate suitability and space requirements. With hundreds of breeds to choose from, each with its unique characteristics and attributes, finding the perfect fit for your needs can seem like a daunting task. However, with a bit of research and consideration, you can narrow down your options and select the breed that best suits your lifestyle, environment, and goals.

One of the first considerations when choosing a chicken breed is egg production. If your primary goal is to have a steady supply of fresh eggs for your family or to sell at the market, you'll want to focus on breeds known for their prolific laying abilities. Popular egg-laying breeds include the Rhode Island Red, Leghorn, and Australorp, all of which are prized for their high egg production rates and reliable performance. Alternatively, if you're less concerned about egg production and more interested in raising chickens for meat or ornamental purposes, there are plenty of breeds to choose from that excel in these areas as well.

Another important consideration when selecting a breed is temperament. While some breeds are known for their docile and friendly nature, others may be more flighty or aggressive, making them less suitable for novice chicken keepers or households with young children. Breeds such as the Buff Orpington, Plymouth Rock, and Sussex are renowned for their calm and gentle demeanor, making them ideal choices for backyard flocks where handling and interaction are important. On the other hand, breeds like the Gamefowl and Old English Game are prized for their feisty personalities and competitive spirit, making them better suited for enthusiasts interested in poultry shows or cockfighting.

Climate suitability is another crucial factor to consider when choosing a chicken breed. Different breeds have varying levels of cold and heat tolerance, so it's important to select a breed that is well-suited to your local climate conditions. For example, breeds with thick, heavy plumage are better able to withstand cold temperatures, while breeds with lighter, more streamlined bodies are better equipped to handle hot, humid climates. Additionally, breeds that have been bred for specific geographic regions, such as the Brahma for cold northern climates or the Leghorn for hot southern climates, may be better adapted to thrive in those environments.

Space requirements are also an important consideration when choosing a chicken breed. Some breeds are more suited to confinement and can thrive in smaller coop and run setups, while others require more space to roam and forage. If you have limited space available or live in an urban or suburban environment, breeds such as the Silkie, Bantam, or Polish may be better suited to your needs due to their smaller size and less active nature. Alternatively, if you have plenty of space available and are looking to raise chickens for free-ranging or pasture-based production, larger breeds such as the Jersey Giant, Orpington, or Wyandotte may be a better fit.

Understanding Basic Chicken Behavior

Understanding basic chicken behavior is essential for any chicken keeper, whether you're a novice enthusiast or a seasoned farmer. Chickens, like all animals, have their unique behaviors and instincts that govern their interactions with each other and their environment. By gaining insight into these behaviors, you can better care for your flock, anticipate their needs, and foster a harmonious and productive living environment for your chickens.

One of the most fundamental aspects of chicken behavior is their social structure and hierarchy. Chickens are naturally social animals that live in hierarchical groups known as flocks. Within a flock, each chicken occupies a specific rank or pecking order, which determines their access to resources such as food, water, and roosting space. Understanding the pecking order is crucial for maintaining peace and harmony within the flock, as conflicts can arise if the hierarchy is disrupted or challenged.

Another important aspect of chicken behavior is their foraging instincts. Chickens are natural foragers that spend much of their time pecking and scratching at the ground in search of insects, seeds, and other edible items. Providing opportunities for your chickens to engage in natural foraging behavior not only helps to satisfy their innate instincts but also promotes physical and mental stimulation, which is essential for their overall health and well-being.

In addition to foraging, chickens also engage in a variety of other behaviors, including dust bathing, roosting, and preening. Dust bathing is a behavior in which chickens roll around in loose soil or sand, often with their wings spread and feathers fluffed, to clean themselves and remove parasites. Providing a designated dust bathing area in your coop or run encourages this natural behavior and helps to keep your chickens clean and healthy.

Roosting is another important behavior for chickens, especially at night when they sleep. Chickens have an instinct to roost off the ground, usually on elevated perches or roosting bars within the coop. Providing ample roosting space and comfortable perches for your chickens to rest on ensures that they feel safe and secure at night and helps to prevent injuries or stress-related behaviors.

Preening is the act of grooming and maintaining their feathers, which is essential for keeping them clean, healthy, and insulated. Chickens use their beaks to preen and oil glands located near the base of their tail to distribute natural oils throughout their feathers.

Observing your chickens preening is a sign of good health and well-being, as it indicates that they are taking care of their grooming needs.

Finally, it's important to recognize that chickens are intelligent and curious animals that thrive on mental stimulation and enrichment. Providing environmental enrichment such as toys, perches, and opportunities for exploration encourages natural behaviors and prevents boredom and stress-related behaviors such as feather pecking and aggression. By understanding and catering to the behavioral needs of your chickens, you can create a fulfilling and enriching living environment that promotes their overall health and happiness.

Setting Up Your Chicken Coop

Setting up a chicken coop is one of the first and most important steps in establishing a successful chicken-keeping operation. A well-designed and properly equipped coop provides your chickens with a safe, comfortable, and secure living environment where they can thrive and flourish. Whether you're building your coop from scratch or repurposing an existing structure, there are several key factors to consider to ensure that your coop meets the needs of your flock and facilitates their health and well-being.

One of the first considerations when setting up your chicken coop is size and space. Chickens require ample space to move around, stretch their wings, and engage in natural behaviors such as foraging and dust bathing. As a general rule of thumb, allow at least 2-3 square feet of indoor space per chicken and 8-10 square feet of outdoor space per chicken in the run or outdoor enclosure. Providing plenty of space helps to prevent overcrowding and reduces the risk of stress-related behaviors such as aggression and feather pecking.

Another important aspect of coop design is ventilation. Adequate ventilation is essential for maintaining good air quality and preventing respiratory issues in your chickens. Ensure that your coop has plenty of windows, vents, and openings to allow for airflow and circulation, especially during hot and humid weather. Additionally, consider installing hardware cloth or wire mesh over openings to prevent predators from gaining access to the coop while still allowing for ventilation.

Additionally, consider the layout and design of your coop to maximize functionality and ease of maintenance. Features such as nesting boxes, roosting bars, and easy-to-clean flooring materials contribute to a comfortable and practical living space for your chickens. Nesting boxes should be located in a quiet, secluded area of the coop and filled with clean bedding material such as straw or shavings to provide a cozy spot for hens to lay their eggs. Roosting bars should be positioned at varying heights to accommodate the natural pecking order and provide plenty of space for all of your chickens to roost comfortably.

Furthermore, it's important to consider the safety and security of your coop to protect your flock from predators and other potential threats. Ensure that your coop is constructed from sturdy materials and that all doors, windows, and openings are securely fastened to prevent access by predators such as raccoons, foxes, and snakes. Additionally, consider installing a predator-proof perimeter fence or wire mesh around the outdoor enclosure to further deter predators and keep your chickens safe.

Chapter 2: Acquiring Your Chickens

Acquiring your chickens is an exciting milestone in your journey as a chicken keeper, marking the beginning of your flock and the start of a rewarding and fulfilling experience. Whether you're starting with a batch of day-old chicks or introducing adult birds to your flock, the process of acquiring chickens requires careful consideration and preparation to ensure the health, safety, and well-being of your new feathered friends. In this chapter, we'll explore the various options for acquiring chickens, from purchasing day-old chicks to adopting adult birds, and discuss the advantages and considerations associated with each approach.

Buying Day-Old Chicks vs. Adult Birds

One of the first decisions you'll need to make when acquiring chickens is whether to start with day-old chicks or adult birds. Both options have their unique advantages and considerations, and the choice ultimately depends on your preferences, experience level, and specific needs as a chicken keeper.

Starting with day-old chicks offers several advantages, particularly for novice chicken keepers or those looking to raise a specific breed or variety of chicken. Chicks are adorable and endearing creatures that captivate our hearts with their fluffy down and tiny peeps, making them a popular choice for first-time chicken keepers. Additionally, raising chicks from day old allows you to witness the entire growth and development process, from cute, fluffy chicks to mature, egg-laying hens.

Another advantage of starting with day-old chicks is the opportunity to bond with your chickens from a young age and establish a strong relationship based on trust and familiarity. By handling and interacting with your chicks regularly, you can help to socialize them and acclimate them to human contact, which can lead to friendlier and more docile chickens as they mature. Additionally, raising chicks from day old allows you to monitor their health and development closely and intervene if any issues arise, such as illness or injury.

Furthermore, starting with day-old chicks allows you to have greater control over the genetics and lineage of your flock, particularly if you're interested in breeding or showing chickens. By selecting chicks from reputable breeders or hatcheries that specialize in your desired breed or variety, you can ensure that your flock exhibits the desired traits, characteristics, and conformation standards associated with that breed.

However, it's important to consider the challenges and responsibilities associated with raising day-old chicks, particularly in terms of care and management. Chicks require specialized care and attention during the first few weeks of life, including access to heat, proper nutrition, and protection from predators and disease. Additionally, raising chicks requires a significant time investment, as they require frequent feedings, clean bedding, and monitoring to ensure their health and well-being.

Alternatively, acquiring adult birds offers several advantages for chicken keepers who are looking for a more immediate and low-maintenance option. Adult birds are already fully feathered, mature, and accustomed to life in a flock, making them less vulnerable to predation and environmental stressors compared to day-old chicks. Additionally, adult birds are typically already laying eggs, which means you can start enjoying fresh eggs from your flock right away.

Another advantage of acquiring adult birds is the reduced time and effort required for care and management compared to raising day-old chicks. Adult birds are self-sufficient and require minimal supervision and maintenance, allowing you to focus on other aspects of chicken keeping such as coop maintenance, egg collection, and flock management. Additionally, adult birds can serve as mentors and role models for younger birds, helping to establish and maintain social order within the flock.

Furthermore, acquiring adult birds offers the opportunity to rescue or rehome chickens in need of a loving and caring home. Many adult birds are available for adoption through local animal shelters, rescue organizations, or online classifieds, providing an opportunity to give a second chance to chickens in need. By adopting adult birds, you not only provide them with a safe and comfortable home but also contribute to the welfare and well-being of animals in need.

However, it's important to consider the potential challenges and considerations associated with acquiring adult birds, particularly in terms of health and compatibility. Adult birds may come with pre-existing health issues or behavioral problems that require attention and intervention, such as respiratory infections, feather picking, or aggression. Additionally, integrating adult birds into an existing flock can be challenging, as established hierarchies and social dynamics may lead to conflicts and aggression among birds.

Selecting Healthy Chickens

Selecting healthy chickens is essential for building a thriving and resilient flock that will provide you with years of enjoyment and productivity. Whether you're purchasing day-old chicks or adult birds, it's important to choose individuals who are free from illness, injury, and genetic defects to ensure the long-term health and well-being of your flock. In this section, we'll discuss the key factors to consider when selecting chickens and offer practical tips for identifying healthy birds.

One of the first things to look for when selecting chickens is overall appearance and condition. Healthy chickens should have bright, alert eyes; clean, smooth feathers; and smooth, unblemished skin. Avoid birds that appear lethargic, disoriented, or unresponsive, as these may be signs of illness or distress. Additionally, check for any signs of injury, such as cuts, bruises, or missing feathers, which could indicate poor handling or aggressive behavior within the flock.

Another important consideration when selecting chickens is respiratory health. Respiratory infections are common in chickens and can spread quickly within a flock if left untreated. When inspecting birds, listen for any abnormal sounds such as wheezing, coughing, or sneezing, which may indicate respiratory issues. Additionally, check for any discharge from the eyes or nostrils, which could be a sign of infection or illness.

In addition to physical health, it's important to consider the genetic background and lineage of the chickens you're considering. Select birds from reputable breeders or hatcheries that prioritize breeding for health, vitality, and genetic diversity. Avoid birds from sources with a history of inbreeding or poor breeding practices, as this can lead to genetic defects and health problems down the line. Look for birds with strong, well-proportioned bodies and traits that are characteristic of their breed or variety.

Furthermore, consider the age and stage of development of the chickens you're selecting. Day-old chicks should be active, alert, and energetic, with fluffy down and clean, unblemished skin. Avoid chicks that appear weak, lethargic, or unresponsive, as these may be signs of underlying health issues. Adult birds should be fully feathered and mature, with a well-developed body and good muscle tone. Avoid birds that are excessively thin or emaciated, as this may indicate poor nutrition or underlying health problems.

Lastly, consider the environment and conditions in which the chickens were raised. Choose birds that have been raised in clean, sanitary conditions with access to fresh water, nutritious feed, and adequate space to move around and exercise. Avoid birds from sources with a history of overcrowding, poor sanitation, or neglect, as these conditions can compromise the health and well-being of the birds and increase the risk of disease transmission.

In summary, selecting healthy chickens is a critical step in building a strong and resilient flock that will provide you with years of enjoyment and productivity. By considering factors such as overall appearance and condition, respiratory health, genetic background, age and stage of development, and environmental conditions, you can identify birds that are free from illness, injury, and genetic defects and set your flock up for success.

Transporting Your Chickens Safely

Transporting chickens safely is essential for ensuring their health and well-being during transit and minimizing stress and discomfort for the birds. Whether you're bringing home day-old chicks from the hatchery or transporting adult birds to a new location, it's important to take precautions to protect your chickens and ensure a smooth and stress-free journey. In this section, we'll discuss the key considerations and practical tips for transporting chickens safely.

One of the first considerations when transporting chickens is the mode of transportation. For short distances, such as bringing home day-old chicks from the hatchery, a cardboard box or plastic tote with ventilation holes can suffice. For longer distances or larger birds, such as adult chickens or roosters, a sturdy pet carrier or livestock crate with ample ventilation and secure latches is recommended. Choose a transportation container that provides enough space for the chickens to stand, turn around, and lie down comfortably without overcrowding.

Before transporting your chickens, it's important to prepare the transportation container and ensure it's clean, sanitary, and secure. Line the bottom of the container with clean bedding material such as straw or wood shavings to absorb moisture and provide traction for the chickens. Additionally, ensure that the container is well-ventilated to provide fresh air circulation and prevent overheating during transit. Secure the lid or door of the container with sturdy latches or zip ties to prevent escapes and ensure the safety of the birds.

Furthermore, consider the weather and environmental conditions when planning your chicken transport. Avoid transporting chickens during extreme temperatures or inclement weather conditions, as this can increase the risk of heat stress, cold stress, or dehydration for the birds. If transporting chickens during hot weather, provide ample shade and ventilation to prevent overheating, and offer access to water to keep the birds hydrated. Similarly, if transporting chickens during cold weather, provide insulation and protection from drafts to keep the birds warm and comfortable.

When loading chickens into the transportation container, handle them gently and avoid overcrowding to minimize stress and discomfort. Allow each bird enough space to stand, turn around, and lie down comfortably, and avoid stacking or overcrowding birds on top of each other. Additionally, consider separating aggressive or territorial birds to prevent fighting or injury during transit.

During transit, monitor the chickens closely for signs of stress, discomfort, or illness, and take appropriate measures to address any issues that arise. Stop periodically to check on the chickens, offer water and food if needed, and adjust ventilation or temperature controls as necessary to ensure the birds' comfort and well-being. Avoid sudden stops, sharp turns, or rough handling that could jostle or injure the birds, and drive carefully and defensively to minimize the risk of accidents or injuries.

Upon arrival at your destination, carefully unload the chickens from the transportation container and acclimate them to their new environment gradually. Provide access to food, water, and shelter, and allow the birds time to explore and settle into their new surroundings at their own pace. Monitor the chickens closely for signs of stress or illness in the days following transport, and provide any necessary care or intervention to ensure their health and well-being.

Chapter 3: Chicken Nutrition and Feeding

Proper nutrition is essential for the health, vitality, and productivity of your flock. Providing your chickens with a balanced and nutritious diet ensures that they have the energy, strength, and immunity to thrive and flourish in their environment. In this chapter, we'll explore the fundamentals of chicken nutrition and feeding, including understanding their nutritional needs and implementing effective feeding practices to support their overall health and well-being.

Understanding Chicken Nutritional Needs

Chickens, like all animals, require a balanced diet that provides essential nutrients such as protein, carbohydrates, fats, vitamins, and minerals to support their growth, development, and overall health. Understanding the nutritional needs of chickens is essential for ensuring that they receive the appropriate nutrients in the right proportions to meet their physiological requirements and maintain optimal health.

One of the most important nutrients for chickens is protein, which is essential for muscle development, feather production, and overall growth and repair. Chickens require a diet that is rich in high-quality protein sources, such as poultry meal, fish meal, soybean meal, and legumes, to meet their protein requirements. Additionally, protein requirements vary depending on factors such as age, breed, and stage of production, so it's important to provide chickens with a diet that meets their specific needs.

Carbohydrates are another important component of a chicken's diet, providing the energy needed to support their daily activities and metabolic processes. Chickens obtain carbohydrates from grains such as corn, wheat, barley, and oats, as well as from fruits and vegetables. Including a variety of carbohydrate sources in their diet ensures that chickens receive a balanced and nutritious supply of energy to fuel their daily activities and maintain optimal health.

Fats are essential for chickens' health and well-being, serving as a concentrated source of energy and providing essential fatty acids that support cellular function, hormone production, and immune function. Including fats in their diet helps to ensure that chickens receive the energy and nutrients they need to maintain optimal health and productivity. Common sources of fats in a chicken's diet include vegetable oils, animal fats, and seeds.

Vitamins and minerals play a crucial role in supporting the overall health and well-being of chickens, serving as cofactors for enzyme reactions, antioxidants, and structural components of cells and tissues. Chickens require a variety of vitamins and minerals, including vitamin A, vitamin D, vitamin E, calcium, phosphorus, and selenium, to support their growth, development, and immune function. Providing chickens with a balanced and varied diet that includes a wide range of vitamin and mineral sources helps to ensure that they receive the nutrients they need to thrive.

In addition to macronutrients and micronutrients, chickens also require access to clean, fresh water at all times to support their hydration, digestion, and overall health. Water plays a crucial role in maintaining chickens' body temperature, transporting nutrients throughout their bodies, and eliminating waste products from their systems. Providing chickens with access to clean, fresh water ensures that they remain hydrated and healthy, especially during hot weather or periods of increased activity.

When formulating a diet for your chickens, it's important to consider their specific nutritional needs based on factors such as age, breed, stage of production, and environmental conditions. For example, growing chicks require a diet that is higher in protein to support their rapid growth and development, while laying hens require a diet that is higher in calcium to support egg production and shell quality. Additionally, consider any dietary restrictions or special requirements that may apply to your flock, such as allergies or sensitivities to certain ingredients.

Choosing the Right Feed

Selecting the right feed for your chickens is crucial for ensuring their health, productivity, and overall well-being. With a wide variety of feeds available on the market, ranging from commercial pellets and crumbles to whole grains and supplements, choosing the right feed can seem like a daunting task. However, by understanding the nutritional needs of your flock and considering factors such as age, breed, stage of production, and environmental conditions, you can select a feed that meets their specific requirements and supports their optimal health and performance.

One of the first considerations when choosing chicken feed is the age and stage of development of your flock. Chickens have different nutritional requirements at various stages of their lives, so it's important to select a feed that is specifically formulated to meet their needs. For example, starter feeds are designed for young chicks and provide the high levels of protein and essential nutrients needed to support their rapid growth and development. Grower feeds are formulated for adolescent chickens and provide a balanced diet that promotes healthy muscle and skeletal development. Layer feeds are designed for laying hens and provide the calcium and other nutrients needed to support egg production and shell quality.

Additionally, consider the breed and size of your chickens when selecting feed. Different breeds and sizes of chickens have different nutritional requirements, so it's important to choose a feed that is appropriate for your specific flock. For example, larger breeds such as Orpingtons and Jersey Giants may require feeds with higher protein and energy content to support their larger size and higher metabolism, while smaller breeds such as Bantams may require feeds with smaller particle sizes and higher nutrient density to meet their specific needs.

Another important consideration when choosing chicken feed is the quality and composition of the ingredients. Look for feeds that are made from high-quality, natural ingredients and free from artificial additives, preservatives, and fillers. Ingredients such as whole grains, protein-rich meals, vitamins, and minerals provide the essential nutrients needed to support chickens' health and well-being. Additionally, consider the sourcing and production practices of the feed manufacturer, and choose feeds from reputable companies with a track record of quality and integrity.

Furthermore, consider any dietary restrictions or special requirements that may apply to your flock when selecting feed. For example, some chickens may have allergies or sensitivities to certain ingredients, such as soy or corn, which may require specialized feeds or alternative sources of protein. Additionally, consider any health issues or conditions that may affect your flock, such as molting, brooding, or illness, and choose feeds that are appropriate for their specific needs.

Supplementing with Treats and Scrap

Supplementing your chickens' diet with treats and scraps is a popular and enjoyable way to provide variety and enrichment to their diet while also reducing waste and saving money on feed costs. Treats and scraps can include a wide variety of foods, including fruits, vegetables, grains, and protein sources, and can provide additional nutrients, energy, and entertainment for your flock. However, it's important to offer treats and scraps in moderation and to choose foods that are safe and appropriate for chickens to consume.

One of the most popular treats for chickens is kitchen scraps, including fruits and vegetables that are past their prime or leftovers from meal preparation. Chickens enjoy a wide variety of fruits and vegetables, including apples, bananas, carrots, cucumbers, lettuce, spinach, and squash, among others. These foods not only provide additional nutrients, vitamins, and minerals to their diet but also provide entertainment and enrichment as chickens peck and scratch at the food.

Additionally, consider offering grains and seeds as treats for your chickens, which provide energy and essential nutrients such as protein, carbohydrates, and fats. Grains such as corn, oats, barley, and wheat are popular treats for chickens and can be offered whole or cracked for easy consumption. Additionally, seeds such as sunflower seeds, pumpkin seeds, and flaxseeds provide additional protein and healthy fats and can be sprinkled on top of feed or offered as a standalone treat.

Furthermore, consider offering protein-rich treats such as mealworms, earthworms, or cooked eggs to your chickens, which provide essential amino acids and support muscle development, feather production, and overall health. Mealworms, in particular, are a favorite treat for chickens and can be purchased dried or live from pet stores or online retailers. Additionally, consider offering cooked eggs, which provide a highly digestible source of protein and essential nutrients and can be scrambled, boiled, or offered in other forms.

When offering treats and scraps to your chickens, it's important to do so in moderation and to avoid foods that are high in salt, sugar, or fat, as these can be harmful to their health. Additionally, avoid offering foods that are toxic or harmful to chickens, such as chocolate, caffeine, avocado, onions, garlic, and raw potatoes, among others. Be mindful of any allergies or sensitivities that your chickens may have and avoid offering foods that could trigger an adverse reaction. Introduce new treats gradually and observe your chickens' response to ensure that they tolerate them well and do not experience any digestive upset or other adverse effects.

In addition to offering treats and scraps, consider supplementing your chickens' diet with commercial treats and supplements specifically formulated for poultry. These products often contain a balanced blend of vitamins, minerals, and other nutrients designed to support chickens' health and well-being. Look for treats and supplements that are made from high-quality, natural ingredients and free from artificial additives, preservatives, and fillers. Offer treats and supplements as directed by the manufacturer, and be mindful of any dietary restrictions or special requirements that may apply to your flock.

When supplementing with treats and scraps, it's important to consider the nutritional content and balance of your chickens' overall diet. Treats and scraps should complement, rather than replace, their regular feed, and should be offered in moderation to prevent nutritional imbalances or deficiencies. Aim to provide treats and scraps as a supplement to their diet, rather than as a primary source of nutrition, and offer a wide variety of foods to ensure that your chickens receive a balanced and nutritious diet.

Furthermore, consider the impact of treats and scraps on your chickens' behavior and social dynamics within the flock. Offering treats can be a fun way to interact with your chickens and strengthen your bond with them. However, be mindful of any competition or aggression that may arise among flock members when offering treats, and consider spreading treats out evenly or providing multiple feeding stations to prevent conflicts and ensure that all chickens have access to treats.

Chapter 4: Health and Wellness

Maintaining the health and wellness of your flock is essential for ensuring their longevity, productivity, and overall well-being. As a responsible chicken keeper, it's important to be proactive in caring for your chickens and to implement preventative measures to protect them from illness, injury, and disease. In this chapter, we'll explore the key aspects of health and wellness for chickens, including preventative care strategies to keep your flock healthy and thriving.

Preventative Care: Keeping Your Chickens Healthy

Preventative care is the foundation of a healthy and resilient flock and involves implementing proactive measures to prevent illness, injury, and disease before they occur. By establishing a routine of care and implementing effective preventative measures, you can help safeguard the health and well-being of your chickens and minimize the risk of health issues and emergencies.

One of the most important aspects of preventative care for chickens is maintaining a clean and sanitary living environment. Cleanliness is essential for preventing the spread of disease and parasites within the flock and minimizing the risk of illness and infection. Establish a regular schedule for cleaning and disinfecting your chicken coop, including removing soiled bedding, scrubbing surfaces with a mild detergent, and disinfecting with a poultry-safe disinfectant. Additionally, keep the surrounding area clean and free from debris, standing water, and other potential sources of contamination.

Another key aspect of preventative care is providing your chickens with a balanced and nutritious diet that meets their specific nutritional needs. A well-balanced diet supports overall health and immunity and helps to prevent nutritional deficiencies and imbalances that can lead to health issues. Ensure that your chickens have access to fresh, clean water at all times, and offer a high-quality feed that is formulated to meet their specific nutritional requirements based on factors such as age, breed, and stage of production.

Additionally, implement a routine of monitoring and observation to detect any signs of illness or injury early and intervene promptly. Spend time observing your chickens daily, paying attention to their behavior, posture, and appearance, and be alert for any changes or abnormalities that may indicate a health issue. Common signs of illness or injury in chickens include lethargy, decreased appetite, respiratory symptoms, abnormal droppings, changes in posture or movement, and abnormal feathering or skin conditions.

Furthermore, practice good biosecurity measures to prevent the introduction and spread of disease within your flock. Biosecurity involves implementing measures to minimize the risk of disease transmission from external sources, such as wild birds, rodents, and other animals, as well as from visitors, equipment, and other flocks. Implement measures such as limiting visitor access to your flock, quarantining new birds before introducing them to your existing flock, and practicing proper hygiene and sanitation when handling chickens or their equipment.

Regularly inspect your chickens for signs of external parasites such as lice, mites, and fleas, and implement appropriate measures to control infestations and prevent their spread within the flock. Provide your chickens with dust bathing areas and access to diatomaceous earth or other natural pest control methods to help keep parasites at bay. Additionally, consider implementing a regular parasite prevention program using poultry-safe treatments such as dusts, sprays, or herbal remedies to help keep your flock healthy and comfortable.

Identifying Common Health Issues

As a chicken keeper, it's essential to be able to identify common health issues that may affect your flock so that you can take prompt action to address them and prevent further complications. While chickens are generally hardy and resilient creatures, they are susceptible to a variety of health problems, ranging from minor ailments to more serious conditions. In this section, we'll explore some of the most common health issues that may affect chickens and discuss how to recognize and manage them effectively.

One of the most common health issues that chickens may experience is respiratory illness. Respiratory infections are often caused by bacteria, viruses, or environmental factors such as poor ventilation or high humidity, and can manifest as symptoms such as sneezing, coughing, wheezing, nasal discharge, and difficulty breathing. In severe cases, respiratory infections can lead to pneumonia or other respiratory complications. To prevent respiratory illness, ensure that your chicken coop is well-ventilated, clean, and free from drafts, and provide your chickens with a clean, dry bedding material to minimize exposure to pathogens.

Another common health issue in chickens is parasitic infestation. External parasites such as lice, mites, and fleas can cause irritation, discomfort, and skin lesions in chickens, leading to symptoms such as feather loss, skin irritation, and abnormal behavior. Additionally, internal parasites such as worms can affect chickens' gastrointestinal tract and impair their digestion and nutrient absorption, leading to poor growth, weight loss, and decreased egg production. To prevent parasitic infestations, implement regular parasite prevention measures such as cleaning and disinfecting the coop, providing dust bathing areas, and administering poultry-safe parasite treatments as needed.

Nutritional deficiencies are another common health issue that may affect chickens, particularly if they are not receiving a balanced and nutritious diet. Deficiencies in essential nutrients such as protein, vitamins, and minerals can lead to a variety of health problems, including poor growth, weak immune function, and reproductive issues. To prevent nutritional deficiencies, ensure that your chickens have access to high-quality feed that is formulated to meet their specific nutritional requirements based on factors such as age, breed, and stage of production. Additionally, supplementing their diet with fresh fruits, vegetables, and protein sources can help to ensure that they receive a balanced and varied diet.

Infectious diseases are a significant concern for chicken keepers, as they can spread quickly within a flock and cause serious illness or death. Common infectious diseases in chickens include Newcastle disease, infectious bronchitis, Marek's disease, and avian influenza, among others. Symptoms of infectious diseases may vary depending on the specific disease and may include lethargy, decreased appetite, respiratory symptoms, diarrhea, and sudden death. To prevent infectious diseases, practice good biosecurity measures such as limiting visitor access to your flock, quarantining new birds before introducing them to your existing flock, and practicing proper hygiene and sanitation when handling chickens or their equipment.

Injuries are another common health issue that chickens may experience, particularly if they are kept in overcrowded or unsafe conditions. Common injuries in chickens include cuts, bruises, broken bones, and sprains, which can result from fighting, predator attacks, falls, or other accidents. To prevent injuries, provide your chickens with a safe and secure living environment, free from hazards such as sharp objects, slippery surfaces, or aggressive flock members. Additionally, handle your chickens gently and avoid overcrowding or rough handling that could lead to injuries.

Administering First Aid to Sick or Injured Chickens

Administering first aid to sick or injured chickens is essential for providing immediate relief and support until professional veterinary care can be obtained. While chickens are generally resilient creatures, they may require assistance in certain situations, such as injuries, illness, or emergencies. In this section, we'll discuss some basic first-aid techniques that you can use to help your chickens in times of need.

One of the first steps in administering first aid to a sick or injured chicken is to assess the situation and identify the nature and severity of the problem. Take the time to observe your chicken closely and assess their symptoms, behavior, and overall condition. Look for signs of injury, illness, or distress, such as limping, lethargy, decreased appetite, abnormal breathing, or abnormal posture. Additionally, consider any recent changes or events that may have contributed to the problem, such as predator attacks, extreme weather conditions, or changes in diet or environment.

Once you have assessed the situation, take appropriate action to address the problem and provide immediate relief to your chicken. For minor injuries such as cuts, scrapes, or bruises, clean the affected area with a mild antiseptic solution and apply a topical antibiotic ointment or wound spray to prevent infection and promote healing. For more serious injuries such as broken bones or deep wounds, immobilize the affected limb or body part and seek veterinary care as soon as possible.

In cases of illness or respiratory distress, isolate the affected chicken from the rest of the flock to prevent the spread of disease and provide supportive care such as warmth, hydration, and nutritional support. Keep the chicken in a quiet, stress-free environment and offer fresh water, electrolytes, and nutritious foods such as scrambled eggs, yogurt, or high-protein treats to help boost their immune system and support recovery.

Additionally, consider implementing supportive measures such as providing heat lamps or heating pads to maintain body temperature, offering supplemental oxygen to support respiratory function, and administering oral or injectable medications as directed by a veterinarian. Keep your chicken comfortable and well-supported throughout the recovery process, and monitor their progress closely for any signs of improvement or deterioration.

In cases of emergencies such as severe injury, illness, or egg binding, seek veterinary care immediately to ensure that your chicken receives prompt and appropriate treatment. Keep a first aid kit stocked with essential supplies such as bandages, antiseptics, wound dressings, and medications, and familiarize yourself with basic first aid techniques for chickens. By being prepared and proactive, you can provide effective first aid to your chickens in times of need and help to ensure their health, safety, and well-being.

Chapter 5: Egg Production and Care

Egg production is a central aspect of chicken keeping, whether for personal consumption, sale, or breeding purposes. Understanding the egg-laying process and providing appropriate care for laying hens is essential for ensuring a consistent supply of high-quality eggs and maintaining the health and well-being of your flock. In this chapter, we'll explore the intricacies of egg production and care, delving into the various stages of the egg-laying process and discussing how to support your hens for optimal egg production and quality.

Understanding the Egg-Laying Process

The egg-laying process is a complex physiological phenomenon that occurs within the reproductive system of hens. It begins with the maturation and release of an ovum, or egg cell, from the hen's ovary, followed by the passage of the egg through the oviduct, where it undergoes fertilization (if mating has occurred) and the formation of the eggshell, membranes, and other protective structures. Finally, the fully formed egg is laid by the hen and deposited into the nest or laying box for collection.

The egg-laying process begins with the development and maturation of ovarian follicles within the hen's ovary. Each follicle contains an ovum, or egg cell, surrounded by layers of protective cells and fluids. As the follicles mature, they release mature eggs into the oviduct, where they begin their journey toward becoming fully formed eggs.

Once released from the ovary, the egg travels through the oviduct, a long, convoluted tube that extends from the ovary to the cloaca, or vent, at the base of the hen's abdomen. Along the way, the egg passes through various sections of the oviduct, each of which plays a specific role in the formation and development of the egg.

The first section of the oviduct, known as the infundibulum, is where fertilization occurs (if mating has occurred). If a rooster has successfully mated with the hen, sperm cells will be present in the infundibulum to fertilize the egg as it passes through. If fertilization does not occur, the egg continues its journey through the oviduct unfertilized.

After passing through the infundibulum, the egg enters the magnum, the longest section of the oviduct, where the egg white, or albumen, is deposited around the yolk. The egg white provides protection and nutrients for the developing embryo and helps to cushion and support the yolk within the egg.

Next, the egg enters the isthmus, where the shell membranes are formed and the eggshell begins to take shape. The shell membranes provide additional protection and support for the developing embryo and help to maintain the structural integrity of the egg.

Finally, the egg enters the uterus, or shell gland, where the eggshell is formed. The shell gland secretes calcium carbonate and other minerals onto the surface of the egg, forming the hard, protective shell that encases the egg. The egg spends the longest amount of time in the uterus, allowing the shell to fully develop and harden before being laid.

Once the egg is fully formed, it is laid by the hen and deposited into the nest or laying box for collection. The entire egg-laying process takes approximately 24-26 hours to complete, with most hens laying eggs in the early morning hours.

Collecting and Handling Egg

Collecting and handling eggs properly is crucial for maintaining their quality, freshness, and safety. Whether you raise chickens for personal consumption or commercial purposes, following best practices for egg collection and handling ensures that you can enjoy delicious and nutritious eggs while minimizing the risk of contamination and spoilage. In this section, we'll explore the steps involved in collecting and handling eggs, from nest inspection to storage, and discuss how to maintain the integrity of the eggs throughout the process.

The first step in collecting eggs is to establish a routine for nest inspection and egg collection. Depending on the size of your flock and the layout of your coop, this may involve checking nests multiple times per day to ensure that eggs are collected promptly after laying. Inspect each nest thoroughly for any signs of damage, contamination, or broodiness, and remove any soiled bedding or debris to maintain a clean and sanitary environment for egg laying.

When collecting eggs, it's important to handle them with care to prevent damage and ensure their quality. Gently lift each egg from the nest, being careful not to jostle or shake it, and place it into a clean and sturdy egg basket or container. Avoid dropping or bumping the eggs against hard surfaces, as this can cause cracks or breakages that compromise their integrity and safety.

Once you have collected all the eggs from the nests, inspect them carefully for any signs of damage, contamination, or irregularities. Discard any eggs that are cracked, dirty, or soiled, as these may harbor bacteria or pathogens that can contaminate the rest of the eggs. Additionally, check for abnormalities such as misshapen eggs or thin shells, which may indicate underlying health issues or nutritional deficiencies in the hens.

After inspecting the eggs, it's important to clean and sanitize them before storage to remove any dirt, debris, or bacteria that may be present on the surface. While eggs are naturally equipped with protective coatings that help to keep out contaminants, they can still become soiled during the laying process or from contact with dirty nesting materials. To clean eggs, gently wipe them with a damp cloth or sponge, taking care not to scrub or rub them too vigorously, as this can damage the protective cuticle.

Once the eggs have been cleaned, they should be promptly refrigerated to maintain their freshness and safety. Refrigeration slows down the growth of bacteria and helps to extend the shelf life of eggs, ensuring that they remain safe and wholesome for consumption. Store eggs in the main body of the refrigerator, rather than in the door, where temperatures may fluctuate more widely, and keep them in their original carton to protect them from absorbing odors and flavors from other foods.

When handling and storing eggs, it's important to practice good food safety habits to minimize the risk of contamination and foodborne illness. Wash your hands thoroughly before and after handling eggs, and avoid cross-contamination by keeping raw eggs separate from ready-to-eat foods and other perishable items. Additionally, cook eggs thoroughly before consuming them, especially if they will be served to young children, pregnant women, or individuals with compromised immune systems.

Maintaining Egg Quality

Maintaining egg quality is essential for ensuring that you can enjoy delicious and nutritious eggs that are free from defects, spoilage, and contamination. Whether you raise chickens for personal consumption or commercial purposes, following best practices for egg handling, storage, and preparation helps to preserve the freshness and integrity of the eggs and ensures a superior eating experience. In this section, we'll explore some key factors that contribute to egg quality and discuss how to maintain it throughout the egg production and handling process.

One of the most important factors in maintaining egg quality is proper handling and storage. Eggs should be handled gently to prevent damage and breakage, as cracks or leaks in the shell can allow bacteria to enter and contaminate the egg. After collecting eggs from the nest, they should be promptly cleaned and sanitized to remove any dirt, debris, or bacteria that may be present on the surface. Avoid washing eggs with soap or detergent, as this can strip away the natural protective coating and increase the risk of contamination.

Once cleaned, eggs should be stored promptly in the refrigerator to maintain their freshness and safety. Refrigeration slows down the growth of bacteria and helps to extend the shelf life of eggs, ensuring that they remain safe and wholesome for consumption. Store eggs in their original carton to protect them from absorbing odors and flavors from other foods, and avoid storing them in the door of the refrigerator, where temperatures may fluctuate more widely.

Another important factor in maintaining egg quality is proper nutrition for the laying hens. Chickens require a balanced diet that provides all the essential nutrients they need to produce high-quality eggs with strong shells, vibrant yolks, and rich flavor. Ensure that your chickens have access to high-quality feed that is formulated to meet their specific nutritional requirements, and supplement their diet with fresh fruits, vegetables, and protein sources to provide additional vitamins and minerals.

In addition to nutrition, environmental factors such as temperature, humidity, and ventilation can also affect egg quality. Keep your chicken coop clean, dry, and well-ventilated to provide a comfortable and stress-free environment for your hens, which in turn promotes healthy egg production. Monitor environmental conditions regularly and make adjustments as needed to ensure that they remain within optimal ranges for egg production and quality.

Proper egg handling and preparation techniques are also important for maintaining egg quality throughout the cooking process. Store eggs in the refrigerator until ready to use, and crack them open on a clean, flat surface to prevent shell fragments from contaminating the egg. Cook eggs thoroughly before consuming them, especially if they will be served to young children, pregnant women, or individuals with compromised immune systems, to reduce the risk of foodborne illness.

Chapter 6: Managing Your Flock

Managing your flock of chickens involves a variety of tasks and responsibilities aimed at ensuring the health, well-being, and productivity of your birds. From handling and socializing your chickens to maintaining a clean and organized coop, effective flock management is essential for fostering a harmonious and thriving environment for your feathered friends. In this chapter, we'll explore the key aspects of managing your flock, including handling and socializing your chickens to promote trust and companionship within the flock.

Handling and Socializing Your Chickens

Handling and socializing your chickens is an important aspect of flock management that helps to promote trust, cooperation, and companionship within the flock. By interacting with your chickens regularly and handling them gently and confidently, you can help to reduce stress, fear, and aggression, and foster positive relationships between flock members and between humans and chickens.

When handling chickens, it's important to approach them calmly and quietly to avoid startling or frightening them. Move slowly and deliberately, and avoid making sudden movements or loud noises that may cause your chickens to become nervous or defensive. Speak to your chickens in a soothing tone of voice and offer treats or rewards to encourage them to approach you willingly and trustingly.

When picking up or holding chickens, support their body securely with both hands, being careful not to squeeze or grasp them too tightly. Use a gentle but firm grip to prevent them from flapping their wings or struggling, and avoid restraining them for longer than necessary to minimize stress and discomfort. Hold chickens close to your body to provide them with a sense of security and stability, and avoid dangling or swinging them, as this can cause distress and anxiety.

Socializing your chickens involves spending time with them regularly and engaging in activities that encourage positive interactions and bonding. Spend time in the chicken coop or run each day, interacting with your chickens, observing their behavior, and providing enrichment activities such as dust bathing areas, perches, and toys to keep them entertained and stimulated.

Encourage socialization and cooperation within the flock by providing ample space, resources, and opportunities for chickens to interact and establish social hierarchies. Introduce new birds to the flock gradually and observe their interactions closely to ensure that they are accepted and integrated smoothly. Monitor flock dynamics and intervene as needed to prevent bullying, aggression, or other negative behaviors that may disrupt the harmony of the flock.

In addition to handling and socializing your chickens, it's important to provide them with a clean, safe, and comfortable living environment that meets their physical and behavioral needs. Keep the chicken coop clean and well-maintained, removing soiled bedding, debris, and waste regularly to prevent the buildup of ammonia and other harmful substances that can affect respiratory health.

Provide ample space for your chickens to roam and forage, both indoors and outdoors, to promote physical activity, mental stimulation, and natural behaviors such as scratching, pecking, and dust bathing. Ensure that the coop is adequately ventilated to provide fresh air and prevent the buildup of moisture, humidity, and odors that can contribute to respiratory problems and other health issues.

Offer a balanced and nutritious diet that meets the specific nutritional requirements of your chickens based on factors such as age, breed, and stage of production. Provide access to fresh, clean water at all times, and offer supplemental feed, treats, and forage to provide variety and enrichment to their diet.

Dealing with Behavioral Issues

Behavioral issues among chickens can arise for various reasons and can impact the harmony and well-being of the flock. From aggression and bullying to egg eating and feather picking, understanding and addressing these behavioral issues is essential for maintaining a peaceful and productive flock environment. In this section, we'll explore some common behavioral issues in chickens and discuss strategies for managing and resolving them effectively.

Aggression and bullying are common behavioral issues among chickens, particularly when introducing new birds to an existing flock or during periods of stress or overcrowding. Aggressive behaviors may include pecking, chasing, fighting, and feather plucking, and can result in injury, stress, and social disruption within the flock. To address aggression and bullying, it's important to identify the underlying causes and implement appropriate interventions to promote peaceful coexistence among flock members.

One strategy for managing aggression and bullying is to provide ample space, resources, and opportunities for chickens to establish social hierarchies and establish territories within the flock. Ensure that the chicken coop and run are adequately sized to accommodate the needs of the flock, and provide multiple feeding and watering stations to prevent competition and reduce conflicts over resources. Additionally, provide hiding spots, perches, and other enrichment activities to allow chickens to escape or avoid confrontations with more dominant flock members.

Another approach to managing aggression and bullying is to observe flock dynamics closely and intervene as needed to prevent or diffuse conflicts before they escalate. Monitor chickens for signs of stress, fear, or aggression, such as raised hackles, flattened combs, or aggressive posturing, and separate birds that are causing problems from the rest of the flock temporarily until tensions subside. Consider using visual barriers or temporary partitions to create separate areas within the coop or run where chickens can be isolated from one another while still being able to see and hear each other.

Egg eating is another common behavioral issue among chickens that can be challenging to address. Egg eating may occur as a result of boredom, nutritional deficiencies, or environmental stressors, and can quickly become a habit that is difficult to break. To prevent egg eating, it's important to identify and address the underlying causes and implement strategies to discourage the behavior and protect the integrity of the eggs.

One strategy for preventing egg eating is to collect eggs promptly after they are laid to minimize the opportunity for chickens to discover and consume them. Establish a routine for nest inspection and egg collection, checking nests multiple times per day if possible, and removing eggs promptly to prevent them from becoming targets for egg-eating behavior. Additionally, provide ample nesting materials such as straw, hay, or shredded paper to encourage chickens to lay eggs in designated nest boxes rather than on the ground where they may be more vulnerable to predation or scavenging.

Feather picking is another common behavioral issue among chickens that can result in injury, stress, and social disruption within the flock. Feather picking may occur as a result of boredom, nutritional deficiencies, overcrowding, or social hierarchy issues, and can quickly escalate if not addressed promptly. To prevent feather picking, it's important to identify and address the underlying causes and implement strategies to promote healthy feather growth and discourage pecking behavior.

One strategy for managing feather picking is to provide ample space, resources, and opportunities for chickens to engage in natural behaviors such as foraging, scratching, and dust bathing. Ensure that the chicken coop and run are adequately sized to accommodate the needs of the flock, and provide plenty of enrichment activities such as perches, dust bathing areas, and hanging toys to keep chickens entertained and stimulated. Additionally, provide a balanced and nutritious diet that meets the specific nutritional requirements of your chickens, and supplement their diet with fresh fruits, vegetables, and protein sources to prevent nutritional deficiencies that may contribute to feather-picking behavior.

Integrating New Birds into the Flock

Integrating new birds into an existing flock can be a challenging process that requires careful planning, patience, and observation. Whether you're adding new chicks, pullets, or adult birds to your flock, it's important to take steps to ensure a smooth and successful transition and minimize stress, aggression, and social disruption among flock members. In this section, we'll explore some strategies for integrating new birds into the flock effectively and promoting harmonious coexistence among all members.

One of the most important aspects of integrating new birds into the flock is to introduce them gradually and in a controlled manner to minimize stress and reduce the risk of aggression or conflict. Begin by placing the new birds in a separate area adjacent to the existing flock, where they can see and hear each other but are not able to make physical contact. This allows the birds to become accustomed to each other's presence and establish a sense of familiarity before being introduced directly.

Once the new birds have had time to acclimate to their new surroundings, gradually introduce them to the existing flock under supervised conditions. Start by allowing them to interact through a wire barrier or mesh partition that prevents physical contact but allows them to see and smell each other. Observe their behavior closely for signs of aggression, fear, or stress, and be prepared to intervene if necessary to prevent injuries or conflicts.

As the birds become more comfortable with each other, gradually increase their interaction time and reduce the barriers separating them until they can be safely integrated into the flock. Monitor their interactions closely for any signs of aggression, bullying, or social disruption, and be prepared to separate birds that are causing problems or being overly aggressive. It may take some time for the new birds to establish their place within the existing social hierarchy of the flock, so be patient and allow them to adjust at their own pace.

To facilitate the integration process, provide plenty of space, resources, and opportunities for the birds to establish territories and social hierarchies within the flock. Ensure that the chicken coop and run are large enough to accommodate the needs of all flock members and provide multiple feeding and watering stations to prevent competition and reduce conflicts over resources. Additionally, provides plenty of hiding spots, perches, and other enrichment activities to allow birds to escape or avoid confrontations with more dominant flock members.

It's also important to consider the age, size, and temperament of the new birds when integrating them into the flock. Young chicks and pullets may be more vulnerable to aggression and bullying from older, more established flock members, so take extra precautions to protect them during the integration process. Provide separate feeding and watering stations for the new birds initially to prevent them from being bullied or excluded from access to food and water, and ensure that they have plenty of space to retreat to if they feel threatened or overwhelmed.

In some cases, it may be necessary to separate particularly aggressive or dominant birds from the flock temporarily to allow the new birds to establish themselves and integrate more smoothly. Use visual barriers or temporary partitions to create separate areas within the coop or run where aggressive birds can be isolated from the rest of the flock while still being able to see and hear each other. Monitor their behavior closely and reintroduce them to the flock gradually once tensions have subsided and they can coexist peacefully.

Overall, integrating new birds into the flock requires careful planning, patience, and observation to ensure a smooth and successful transition. By introducing new birds gradually and in a controlled manner, providing plenty of space, resources, and opportunities for social interaction, and monitoring their behavior closely for signs of aggression or stress, you can help to promote harmonious coexistence among all members of the flock and create a positive and supportive environment for your chickens to thrive.

Chapter 7: Coop Maintenance and Cleaning

Proper maintenance and cleaning of the chicken coop are essential tasks for ensuring the health, comfort, and safety of your flock. A clean and well-maintained coop provides a sanitary living environment for your chickens, reduces the risk of disease and pest infestations, and promotes overall flock health and well-being. In this section, we'll explore the importance of coop maintenance and cleaning and discuss best practices for keeping your coop clean, tidy, and hygienic.

Cleaning and Sanitizing the Coop

Regular cleaning and sanitizing of the chicken coop are essential for preventing the buildup of dirt, debris, and harmful bacteria that can pose health risks to your flock. A dirty or unsanitary coop can harbor pathogens, parasites, and other contaminants that can cause respiratory infections, digestive issues, and other health problems in chickens. By implementing a regular cleaning routine and using proper cleaning and sanitizing techniques, you can help maintain a clean and healthy living environment for your birds.

Start by removing all bedding, nesting materials, and debris from the coop, including droppings, feathers, and leftover feed. Use a shovel, rake, or pitchfork to scoop out the soiled bedding and debris, and dispose of it in a compost bin or designated waste area away from the coop. Thoroughly clean and disinfect all surfaces inside the coop, including walls, floors, nesting boxes, roosts, and feeders, using a mild detergent or specialized poultry coop cleaner.

After cleaning, rinse all surfaces thoroughly with clean water to remove any soap residue or cleaning solution. Allow the coop to air dry completely before adding fresh bedding and nesting materials. While the coop is drying, take the opportunity to inspect it for any signs of damage, wear, or structural issues that may need to be addressed, such as loose boards, gaps in the walls or roof, or areas of water damage or rot.

Once the coop is clean and dry, add fresh bedding and nesting materials to provide a clean, comfortable, and sanitary living environment for your chickens. Choose bedding materials such as straw, hay, wood shavings, or shredded paper that are absorbent, dust-free, and non-toxic, and spread them evenly throughout the coop to provide a soft and cushioned surface for your chickens to rest and nest.

In addition to regular cleaning, it's important to sanitize the coop periodically to kill any remaining bacteria, parasites, or pathogens that may be present on surfaces or in the environment. Use a disinfectant solution that is specifically formulated for use in poultry coops, such as a diluted bleach solution or commercial poultry coop sanitizer, and follow the manufacturer's instructions for application and dilution ratios.

Apply the disinfectant solution to all surfaces inside the coop, paying particular attention to areas that are prone to contamination or where bacteria may accumulate, such as feeding and watering stations, nesting boxes, and roosting areas. Allow the disinfectant to sit for the recommended contact time, usually 10-15 minutes, before rinsing thoroughly with clean water and allowing the coop to air dry completely.

In addition to cleaning and sanitizing the coop, it's important to practice good biosecurity measures to prevent the introduction and spread of diseases and pests within your flock. Limit access to the coop to essential personnel only, and avoid sharing equipment, tools, or supplies with other poultry owners to reduce the risk of cross-contamination. Quarantine new birds before introducing them to the flock, and monitor flock health closely for any signs of illness or disease.

Preventing Pest Infestations

Pests pose a significant threat to the health and well-being of your chickens and can cause a range of issues, from transmitting diseases to causing stress and discomfort. Preventing pest infestations is crucial for maintaining a healthy and thriving flock. By implementing proactive measures and practicing good hygiene and management practices, you can minimize the risk of pest problems and ensure the overall well-being of your chickens.

One of the most effective ways to prevent pest infestations in your chicken coop is to maintain a clean and sanitary environment. Regularly clean and disinfect the coop, removing any debris, spilled feed, and accumulated waste that can attract pests such as flies, rodents, and mites. Keep feed and water containers clean and free of spills, and store feed in tightly sealed containers to prevent access by pests.

Another important aspect of pest prevention is to implement proper waste management practices. Remove and properly dispose of any spilled feed, litter, or manure from the coop and surrounding areas regularly to eliminate potential breeding grounds for pests. Consider composting manure in a designated area away from the coop to reduce odors and minimize the attraction of flies and other pests.

Maintaining proper ventilation and airflow in the coop can also help prevent pest infestations by reducing moisture levels and discouraging the growth of mold, mildew, and other pests. Ensure that the coop is well-ventilated with windows, vents, and fans to promote air circulation and prevent the buildup of humidity, which can attract pests such as mites and mold.

Implementing physical barriers and deterrents can also help prevent pests from entering the coop and accessing your chickens. Seal any cracks, gaps, or openings in the coop walls, floors, and roof to prevent entry by rodents, snakes, and other pests. Install wire mesh or hardware cloth around the coop perimeter and openings to prevent access by predators such as raccoons, foxes, and birds of prey.

Regularly inspect the coop and surrounding areas for signs of pest activity, such as droppings, nests, or damage to the coop structure. Be proactive in addressing any pest issues that arise, such as treating mites or lice infestations promptly and sealing any entry points or potential nesting sites for rodents and other pests.

In addition to implementing preventative measures, consider using natural and organic pest control methods to manage pest populations in and around the coop. For example, introduce beneficial insects such as ladybugs, lacewings, and predatory mites to help control populations of harmful pests such as aphids, flies, and mites. Plant pest-repelling herbs and flowers such as lavender, mint, and marigolds around the coop to deter pests naturally.

Regularly monitor your chickens for signs of pest infestations, such as excessive scratching, feather loss, or irritated skin, and take prompt action to address any issues that arise. Inspect your chickens regularly for signs of external parasites such as mites, lice, and fleas, and treat affected birds promptly with appropriate pest control products or remedies.

By implementing proactive measures, practicing good hygiene and management practices, and using natural and organic pest control methods, you can effectively prevent pest infestations in your chicken coop and ensure the health and well-being of your flock.

Routine Maintenance Tasks

Routine maintenance tasks are essential for keeping your chicken coop in good condition and ensuring the health, comfort, and safety of your flock. By staying on top of regular maintenance chores and addressing any issues promptly, you can prevent costly repairs, minimize the risk of pest infestations, and create a clean and comfortable living environment for your chickens.

One of the most important routine maintenance tasks for the chicken coop is to regularly clean and sanitize the coop and surrounding areas. Remove soiled bedding, debris, and waste from the coop run regularly, and clean and disinfect all surfaces inside the coop, including walls, floors, nesting boxes, roosts, and feeders. Use a mild detergent or specialized poultry coop cleaner to clean surfaces, followed by a disinfectant solution to kill any remaining bacteria or pathogens.

In addition to regular cleaning, it's important to inspect the coop and surrounding areas regularly for signs of damage, wear, or structural issues that may need to be addressed. Check for loose or damaged boards, gaps in the walls or roof, and areas of water damage or rot, and repair or replace any damaged or deteriorating components promptly to prevent further issues and maintain the integrity of the coop.

Maintaining proper ventilation and airflow in the coop is also essential for preventing moisture buildup, reducing humidity levels, and promoting air circulation. Ensure that windows, vents, and fans are clean and unobstructed to allow for adequate airflow, and consider installing additional ventilation if needed to improve air quality and prevent respiratory issues in your chickens.

Regularly inspect and clean the coop's nesting boxes, perches, and feeders to ensure that they are in good condition and functioning properly. Replace any worn or damaged nesting materials, perches, or feeders as needed, and provide fresh bedding and nesting materials regularly to keep your chickens comfortable and hygienic.

Inspect the coop's roof and exterior for signs of damage, wear, or deterioration, such as loose or missing shingles, cracks in the siding, or damaged hardware. Repair or replace any damaged or deteriorating components promptly to prevent water leaks, drafts, or pest infestations, and ensure that the coop remains weatherproof and secure.

In addition to regular maintenance tasks, it's important to implement a regular cleaning and disinfection schedule to keep the coop clean and hygienic and prevent the buildup of bacteria, mold, and other contaminants. Plan to clean and disinfect the coop thoroughly at least once per month, or more frequently if needed, and perform spot cleaning and maintenance tasks as needed between deep cleanings.

By staying on top of routine maintenance tasks, you can keep your chicken coop in good condition and ensure the health, comfort, and safety of your flock. By regularly cleaning and sanitizing the coop, inspecting for signs of damage or wear, maintaining proper ventilation and airflow, and addressing any issues promptly, you can create a clean and comfortable living environment for your chickens to thrive.

Chapter 8: Seasonal Considerations

As the seasons change, so do the needs of your chickens. Each season presents its own set of challenges and considerations for chicken care, from extreme temperatures to changes in daylight hours. By understanding the seasonal needs of your flock and implementing appropriate management practices, you can ensure that your chickens stay healthy, comfortable, and productive throughout the year. In this chapter, we'll explore seasonal considerations for chicken keeping, focusing on winter care and how to keep your chickens warm and healthy during the colder months.

Winter Care: Keeping Your Chickens Warm and Healthy

Winter can be a challenging time for chickens, with cold temperatures, snow, ice, and reduced daylight hours posing potential risks to their health and well-being. However, with proper planning and preparation, you can help your chickens weather the winter months comfortably and safely. Here are some tips for winter care to keep your chickens warm and healthy:

1. Provide Adequate Shelter: The first step in winter care for chickens is to ensure that they have adequate shelter from the elements. Make sure that the chicken coop is well-insulated and draft-free, with plenty of bedding to provide insulation and warmth. Consider adding extra insulation to the coop walls and roof, and use weather-stripping or draft guards to seal any gaps or openings that may let cold air in.

2. Maintain Proper Ventilation: While it's important to keep the coop warm during the winter, it's equally important to maintain proper ventilation to prevent the buildup of moisture, ammonia, and other harmful gases. Ensure that the coop has adequate ventilation with windows, vents, and fans to allow for air circulation without creating drafts. Use wire mesh or hardware cloth to cover vents and openings to prevent entry by predators while still allowing for airflow.

3. Provide Heat Sources: In regions with extremely cold temperatures, you may need to provide supplemental heat sources to keep your chickens warm. Consider using heat lamps, radiant heaters, or heated perches to provide warmth inside the coop. Place heat sources safely away from bedding and nesting materials to reduce the risk of fire, and use thermostatically controlled heaters to regulate temperatures and prevent overheating.

4. Offer Warm Water: Chickens need access to clean, unfrozen water at all times, even in the winter. Check waterers frequently and break up any ice that forms to ensure that your chickens have access to water. Consider using heated waterers or placing waterers in a sunny spot to prevent freezing, and add electrolytes or vitamins to the water to help keep your chickens hydrated and healthy.

5. Provide Adequate Nutrition: During the winter months, chickens may require additional calories to maintain their body temperature and energy levels. Ensure that your chickens have access to a balanced and nutritious diet that provides plenty of energy and essential nutrients. Consider increasing their feed rations or supplementing their diet with high-energy treats such as cracked corn, sunflower seeds, or mealworms to help them stay warm and healthy.

6. Protect Against Frostbite: Chickens are susceptible to frostbite on their combs, wattles, and exposed skin in cold weather. To protect against frostbite, apply a thin layer of petroleum jelly or emollient cream to their combs and wattles to help seal in moisture and prevent freezing. Consider installing frostbite guards or providing shelters with low ceilings to minimize exposure to cold temperatures and wind chill.

7. Monitor Health and Behavior: During the winter months, it's important to monitor your chickens closely for signs of cold stress, illness, or injury. Watch for symptoms such as lethargy, reduced appetite, pale combs, or difficulty breathing, and take prompt action if you notice any signs of distress. Provide additional bedding and insulation as needed, and consider bringing sick or injured chickens indoors to recover in a warm and sheltered environment.

8. Adjust Lighting: With shorter daylight hours in the winter, you may need to supplement natural light in the coop to maintain egg production and prevent behavioral issues such as aggression or feather picking. Consider installing artificial lighting on a timer to provide supplemental light in the coop for 14-16 hours per day, simulating longer daylight hours and helping to stimulate egg production.

9. Protect Against Predators: Winter can bring hungry predators looking for an easy meal, so it's important to take extra precautions to protect your flock. Ensure that the coop is secure and predator-proof, with sturdy fencing, locks, and latches to prevent entry by predators such as raccoons, foxes, and coyotes. Consider installing motion-activated lights or alarms to deter nocturnal predators, and remove any fallen branches or debris that may provide cover or hiding spots for predators.

By implementing these tips for winter care, you can help your chickens stay warm, healthy, and comfortable throughout the colder months. With proper planning and preparation, you can ensure that your flock not only survives the winter but thrives, laying eggs, staying active, and maintaining their overall well-being despite the challenges of the season.

Summer Safety: Protecting Chickens from Heat Stress

Summer can be a challenging time for backyard chicken keepers as high temperatures can pose significant risks to the health and well-being of their feathered friends. Heat stress is a serious concern for chickens during the summer months, as they are more susceptible to overheating than many other animals. In this comprehensive guide, we will explore various strategies and techniques for ensuring the safety and comfort of chickens in hot weather.

Understanding Heat Stress in Chickens: Chickens are naturally equipped to handle a wide range of temperatures, but extreme heat can still be detrimental to health. When temperatures rise above a certain threshold, chickens can experience heat stress, which can lead to a range of problems including dehydration, heat stroke, and even death if not properly addressed. Chicken keepers need to understand the signs of heat stress in their flock, which may include panting, lethargy, reduced egg production, and in severe cases, collapse.

Providing Adequate Shade and Ventilation: One of the most important steps in protecting chickens from heat stress is to ensure they have access to adequate shade and ventilation. This can be accomplished by providing shade structures such as awnings, umbrellas, or natural vegetation in their outdoor enclosure. Additionally, ensuring proper ventilation in the chicken coop is essential for allowing hot air to escape and cool air to circulate. This can be achieved by installing windows, vents, or fans to promote airflow.

Hydration and Electrolyte Balance: During hot weather, chickens can quickly become dehydrated if they do not have access to an ample supply of fresh, clean water. It's essential to provide multiple sources of water throughout the chicken enclosure and to regularly check and refill water containers to ensure they do not run dry. Additionally, supplementing their water with electrolytes can help replenish essential nutrients lost through sweating and panting.

Adjusting Feeding and Daily Routine: In extreme heat, chickens may lose their appetite and consume less food than usual. To accommodate this, it may be necessary to adjust their feeding schedule and provide smaller, more frequent meals throughout the day. It's also important to avoid feeding chickens during the hottest part of the day, as this can increase their body temperature and contribute to heat stress. Instead, offering treats such as frozen fruits and vegetables can help keep chickens cool and hydrated.

Monitoring and Intervention: Regular monitoring of chickens during hot weather is crucial for identifying signs of heat stress and taking appropriate action. This may include providing additional shade, misting the area with water to reduce ambient temperature, or bringing chickens indoors to a climate-controlled environment if necessary. It's important to act quickly at the first sign of heat stress to prevent serious health complications and potential loss of life.

Fall and Spring Preparations

As the seasons change from summer to fall and from winter to spring, backyard chicken keepers must prepare their flocks for the transition in weather and environmental conditions. Fall and spring bring their own set of challenges and opportunities for chicken care, from preparing for cooler temperatures and shorter days to ensuring optimal conditions for egg production and overall health. In this comprehensive guide, we will explore various strategies and techniques for preparing chickens for the fall and spring seasons.

Preparing for Cooler Temperatures: As temperatures begin to drop in the fall, it's important to prepare chickens for cooler weather to ensure their health and comfort. This may include providing additional bedding in the coop to help insulate against cold drafts, installing heat lamps or heated waterers to prevent freezing, and adjusting feeding routines to provide extra calories for energy and warmth. It's also essential to regularly inspect the coop for any gaps or leaks that could allow cold air to enter and make necessary repairs.

Managing Daylight Changes: Fall and spring bring changes in daylight hours, which can impact egg production and the overall behavior of chickens. As daylight decreases in the fall, chickens may naturally reduce their egg-laying frequency or even stop laying altogether. To mitigate this, some chicken keepers choose to install artificial lighting in the coop to simulate longer daylight hours and encourage continued egg production. Conversely, as daylight increases in the spring, chickens may experience a surge in egg production, requiring adjustments to feeding and nesting arrangements to accommodate the increased supply.

Protecting Against Predators and Pests: Fall and spring can be prime times for predators and pests to target backyard chicken flocks, as natural food sources become scarce and temperatures fluctuate. Chicken keepers need to remain vigilant and take proactive measures to protect their flock from harm. This may include reinforcing coop security with sturdy locks and barriers, using predator-proof fencing around the enclosure, and implementing pest control measures such as regular cleaning and sanitization of the coop and surrounding area.

Transitioning to Outdoor Living: In the spring, as temperatures begin to warm and daylight hours increase, many chicken keepers choose to transition their flock from indoor confinement to outdoor living. This may involve gradually introducing chickens to outdoor environments, providing access to fresh grass, bugs, and natural sunlight, and monitoring their behavior and health during the transition period. Additionally, ensuring that the outdoor enclosure is secure and free from hazards such as toxic plants or sharp objects is essential for the safety and well-being of the flock.

Chapter 9: Breeding and Incubation

Breeding and incubation are essential aspects of chicken husbandry for those interested in expanding their flock or maintaining specific breeds. Understanding the intricacies of chicken reproduction and the process of incubation is crucial for successfully hatching chicks and ensuring the health and vitality of the next generation.

Understanding Chicken Reproduction

Chicken reproduction is a fascinating and complex process that involves several physiological and behavioral factors. At the core of chicken reproduction is the mating behavior between roosters and hens, which typically occurs when the rooster mounts the hen and transfers sperm to fertilize her eggs. However, successful fertilization also relies on the presence of a viable egg, which is produced by the hen's reproductive system.

The reproductive system of a hen consists of various organs, including the ovary, oviduct, and cloaca. The ovary is responsible for producing ova, or egg cells, which are released into the oviduct for fertilization. Once fertilized, the egg travels through the oviduct, where it undergoes the process of albumen and shell formation before being laid through the cloaca.

The timing of egg production and fertility in chickens is influenced by several factors, including age, breed, nutrition, and environmental conditions. Young hens typically begin laying eggs around 5-6 months of age, although this can vary depending on breed and individual factors. Additionally, certain breeds may have specific reproductive characteristics, such as broodiness or egg color, that can affect breeding and egg production.

Managing a breeding program involves selecting suitable breeding pairs based on desired traits such as size, color, egg production, and temperament. Careful consideration should be given to genetic diversity and avoiding inbreeding, which can lead to undesirable traits and health problems in offspring. Additionally, providing a balanced diet rich in essential nutrients is essential for maintaining reproductive health and fertility in breeding stock.

Incubation Process

Once fertile eggs are obtained, the next step in the breeding process is incubation, which involves artificially maintaining optimal conditions for embryo development until hatching. Incubation can be carried out using various methods, including natural incubation by broody hens or artificial incubation using specialized equipment such as incubators.

Natural incubation occurs when a broody hen expresses a strong maternal instinct to sit on and hatch eggs. During this process, the hen regulates the temperature and humidity of the eggs by sitting on them and occasionally turning them to ensure even heat distribution. Broody hens typically exhibit behaviors such as puffing up their feathers, clucking softly, and refusing to leave the nest.

Artificial incubation, on the other hand, involves using an incubator to recreate the ideal conditions for embryo development. This includes maintaining a constant temperature of around 99.5°F (37.5°C) and humidity levels between 40-50% for the first 18 days of incubation, followed by increased humidity during the final days before hatching. Additionally, eggs should be turned regularly to prevent the embryo from sticking to the shell membrane and ensure proper development.

Successful incubation requires careful monitoring of temperature, humidity, and egg turning throughout the entire process. Deviations from optimal conditions can result in poor hatch rates, developmental abnormalities, and chick mortality. It's also important to handle eggs with care and cleanliness to minimize the risk of contamination and bacterial growth.

Incubating Eggs

Incubating eggs is a crucial step in the process of hatching chicks, whether for commercial purposes or for backyard poultry enthusiasts. The success of incubation relies on maintaining optimal conditions for embryo development, including temperature, humidity, and ventilation, throughout the entire incubation period.

The first step in incubating eggs is to carefully select fertile eggs from healthy breeding stock. Eggs should be collected promptly after being laid and stored in a clean, cool, and humid environment until they are ready for incubation. It's essential to avoid washing or refrigerating eggs, as this can compromise their natural protective coating and reduce hatch rates.

Once ready for incubation, eggs should be placed in a specialized incubator equipped with temperature and humidity controls. The optimal temperature for incubating chicken eggs is around 99.5°F (37.5°C), with slight variations tolerated as long as they do not exceed 1°F (0.5°C) above or below the recommended range for extended periods. Humidity levels should be maintained between 40-50% during the first 18 days of incubation, with a gradual increase to 65-75% during the final days before hatching to facilitate chick emergence.

Proper ventilation is essential for providing a constant supply of fresh air to developing embryos and preventing the buildup of harmful gases such as carbon dioxide. Most modern incubators are equipped with built-in ventilation systems, but it's important to periodically check and adjust airflow as needed to ensure optimal conditions.

Throughout the incubation process, eggs should be regularly turned to prevent the embryo from sticking to the shell membrane and promote even heat distribution. Traditionally, eggs were manually turned several times a day, but many modern incubators feature automatic egg turners that gently rotate eggs at regular intervals. Regardless of the method used, consistency and attention to detail are key to maximizing hatch rates and producing healthy chicks.

Monitoring the development of embryos is another critical aspect of successful egg incubation. Candling, the process of shining a bright light through the egg to visualize internal structures, can be performed at various stages of incubation to assess embryo viability and detect any developmental abnormalities. Eggs that are infertile or contain nonviable embryos should be removed from the incubator to prevent contamination and reduce the risk of spreading disease to viable eggs.

As the incubation period progresses, careful observation of eggs is essential for identifying signs of pipping, the process of chicks breaking through the eggshell to hatch. Pipping typically occurs around day 21 of incubation for chicken eggs, although it can vary depending on breed and environmental factors. Once chicks begin pipping, it's crucial to maintain stable temperature and humidity levels to support the hatching process and ensure the health and vitality of newborn chicks.

Caring for Chicks

Caring for chicks is a critical aspect of poultry husbandry that requires careful attention to their health, nutrition, and environment during the early stages of life. From brooding newborn chicks to transitioning them to outdoor living, proper care and management practices are essential for ensuring their well-being and maximizing their growth and development.

The first step in caring for newborn chicks is to provide them with a warm and comfortable brooding environment to mimic the conditions of a mother hen. A brooder box or pen equipped with a heat source such as a heat lamp or heating pad should be set up before the arrival of chicks, with temperatures maintained between 95-100°F (35-38°C) for the first week of life. It's important to monitor temperature levels regularly and adjust the heat source as needed to prevent chicks from becoming too hot or too cold.

In addition to maintaining optimal temperature, providing clean bedding such as pine shavings or straw is essential for keeping chicks dry and comfortable. Bedding should be changed regularly to prevent the buildup of moisture and ammonia, which can lead to respiratory issues and other health problems. Additionally, providing access to fresh water and chick starter feed formulated specifically for their nutritional needs is crucial for promoting healthy growth and development.

As chicks grow and mature, gradually reducing brooding temperature by 5°F (2-3°C) each week until reaching ambient room temperature is essential for acclimating them to outdoor conditions. This process, known as brooder temperature management, helps chicks develop their thermoregulatory abilities and prepares them for transitioning to outdoor living.

During the brooding period, observing chick behavior and health is essential for identifying any signs of illness or distress. Common health issues in chicks include pasty butt, a condition characterized by the buildup of feces around the vent area, and spraddle leg, a condition where chicks are unable to stand or walk properly due to leg deformities. Prompt intervention and treatment are necessary for addressing these issues and preventing further complications.

As chicks grow and mature, providing opportunities for exercise, exploration, and social interaction is essential for promoting their physical and behavioral development. This may include introducing them to outdoor enclosures equipped with roosts, perches, and enrichment activities such as dust bathing areas and foraging opportunities. Additionally, integrating chicks with older flock members gradually and under supervised conditions can help facilitate socialization and minimize potential conflicts.

Chapter 10: Free-Range and Pasture-Raised Chickens

Benefits of Free-Range and Pasture-Raised Chickens

Free-range and pasture-raised chickens are becoming increasingly popular among consumers who prioritize animal welfare, environmental sustainability, and the quality of food products. Unlike conventionally raised chickens confined to cages or indoor facilities, free-range and pasture-raised chickens have access to outdoor areas where they can roam, forage, and engage in natural behaviors. This chapter explores the numerous benefits associated with free-range and pasture-raised chicken production, ranging from improved animal welfare and nutritional quality to environmental stewardship and sustainable agriculture practices.

Improved Animal Welfare: One of the primary benefits of free-range and pasture-raised chicken production is improved animal welfare. In these systems, chickens are allowed to express natural behaviors such as foraging for food, dust bathing, and socializing with other flock members. Access to outdoor areas with ample space and vegetation promotes physical and psychological well-being, reducing stress and the risk of behavioral problems commonly observed in confined environments. By providing chickens with a more natural and enriching environment, free-range and pasture-raised systems prioritize the health and happiness of the animals, resulting in higher overall welfare standards.

Enhanced Nutritional Quality: In addition to promoting animal welfare, free-range and pasture-raised chicken production can also lead to enhanced nutritional quality in poultry products. Chickens raised in outdoor environments have the opportunity to graze on a diverse array of plants, insects, and other natural food sources, resulting in a diet that is richer in essential nutrients such as vitamins, minerals, and antioxidants. Studies have shown that eggs and meat from free-range and pasture-raised chickens contain higher levels of beneficial nutrients such as omega-3 fatty acids, vitamin E, and beta-carotene compared to conventionally raised counterparts. By consuming products from free-range and pasture-raised chickens, consumers can enjoy the health benefits of nutrient-dense foods while supporting more sustainable farming practices.

Environmental Stewardship: Another significant benefit of free-range and pasture-raised chicken production is its positive impact on the environment. Unlike conventional confinement systems that rely heavily on inputs such as feed, water, and energy, free-range and pasture-based systems utilize natural resources more efficiently and sustainably. Chickens raised in outdoor environments help to control pests and weeds, improve soil fertility through natural fertilization, and contribute to overall ecosystem health and biodiversity. Additionally, rotational grazing practices employed in pasture-based systems can help mitigate soil erosion, promote carbon sequestration, and reduce greenhouse gas emissions, making them more environmentally friendly alternatives to conventional poultry production methods.

Support for Sustainable Agriculture: Free-range and pasture-raised chicken production aligns with the principles of sustainable agriculture by prioritizing environmental stewardship, animal welfare, and community engagement. These systems promote biodiversity, conserve natural resources, and reduce reliance on synthetic inputs such as antibiotics and chemical fertilizers, thereby minimizing the negative impacts of conventional farming practices on ecosystems and public health. Furthermore, supporting local and small-scale producers who practice free-range and pasture-raised chicken production helps to strengthen rural economies, preserve agricultural heritage, and promote food sovereignty and resilience in the face of global challenges such as climate change and food insecurity.

Consumer Demand and Market Opportunities: The growing consumer demand for ethically produced and sustainably sourced food products has led to increased market opportunities for free-range and pasture-raised chicken producers. Consumers are increasingly willing to pay a premium for products that align with their values and preferences, including those that prioritize animal welfare, environmental sustainability, and nutritional quality. As a result, many farmers and producers are transitioning to free-range and pasture-based production systems to meet this demand and capitalize on market opportunities. By offering products that meet consumer expectations for quality, transparency, and ethical production practices, free-range and pasture-raised chicken producers can differentiate themselves in the marketplace and build loyal customer relationships based on trust and integrity.

Considerations for Free-Range Management

Free-range management involves providing chickens with access to outdoor areas where they can roam, forage, and engage in natural behaviors. While free-ranging offers numerous benefits for animal welfare, environmental sustainability, and product quality, it also presents unique challenges and considerations for poultry producers. This section explores key considerations for effectively managing free-range poultry systems, including habitat selection, predator control, flock health, and biosecurity measures.

Habitat Selection: Selecting an appropriate habitat for free-range poultry is essential for ensuring the health and safety of the flock. Ideal habitats should provide ample space for chickens to roam and forage, with access to vegetation, insects, and natural shelter. Pastures, woodlands, and grassy areas with diverse plant species are well-suited for free-range poultry, as they offer a variety of food sources and environmental enrichment opportunities. It's important to assess soil quality, drainage, and vegetation cover when choosing a habitat, as these factors can impact flock health and productivity.

Predator Control: Predator control is a critical aspect of free-range management to protect chickens from potential threats such as foxes, raccoons, hawks, and other predators. Fencing is one of the most effective methods for deterring predators and creating a safe outdoor environment for poultry. Electric fencing, poultry netting, and hardware cloth are commonly used to enclose free-range areas and prevent access by predators. Additionally, installing motion-activated lights, predator deterrents such as scarecrows or reflective tape, and securing coop entrances with predator-proof latches can help minimize the risk of predation.

Flock Health: Maintaining flock health is essential for ensuring the long-term success and productivity of free-range poultry systems. Regular health monitoring, disease prevention, and veterinary care are key components of effective flock management. Providing access to clean water, nutritious feed, and supplemental vitamins and minerals can help support immune function and overall health in free-range chickens. Additionally, practicing good hygiene and sanitation measures, such as regularly cleaning and disinfecting waterers, feeders, and coop bedding, can help prevent the spread of disease and reduce the risk of illness in the flock.

Biosecurity Measures: Implementing biosecurity measures is essential for protecting free-range poultry from infectious diseases and pathogens. Biosecurity protocols should include measures to prevent the introduction of disease agents into the flock, as well as strategies to minimize the spread of disease within and between flocks. This may include restricting access to free-range areas for visitors, vehicles, and other animals, implementing quarantine procedures for new birds or equipment, and practicing strict sanitation and hygiene practices. Additionally, monitoring for signs of illness or disease in the flock and seeking prompt veterinary attention when needed can help prevent outbreaks and mitigate the impact of infectious diseases on flock health and productivity.

Creating a Safe Outdoor Environment

Creating a safe outdoor environment for free-range poultry involves careful planning, management, and implementation of various strategies to ensure the health, safety, and well-being of the flock. This section explores key considerations for creating a safe outdoor environment, including habitat design, predator control, disease prevention, and environmental enrichment.

Habitat Design: Designing a suitable habitat for free-range poultry is essential for providing chickens with access to natural resources and environmental enrichment while minimizing potential risks and hazards. Fencing is a critical component of habitat design to prevent predation and unauthorized access by predators and other animals. Additionally, incorporating natural features such as trees, shrubs, and vegetation can provide shade, shelter, and foraging opportunities for chickens. It's important to consider factors such as soil quality, drainage, and vegetation cover when designing a free-range habitat to ensure optimal conditions for flock health and productivity.

Predator Control: Predator control is a primary concern when creating a safe outdoor environment for free-range poultry. Fencing is an effective method for deterring predators and protecting chickens from potential threats such as foxes, raccoons, and birds of prey. Electric fencing, poultry netting, and hardware cloth are commonly used to enclose free-range areas and prevent access by predators. Additionally, installing motion-activated lights, predator deterrents such as scarecrows or reflective tape, and securing coop entrances with predator-proof latches can help minimize the risk of predation and create a secure outdoor environment for poultry.

Disease Prevention: Preventing the spread of disease is essential for maintaining flock health and productivity in free-range poultry systems. Implementing biosecurity measures such as restricting access to free-range areas for visitors, vehicles, and other animals, practicing good hygiene and sanitation practices, and monitoring for signs of illness or disease in the flock can help minimize the risk of disease transmission. Additionally, vaccination programs, parasite control measures, and regular health monitoring can help prevent the introduction and spread of infectious diseases and pathogens in free-range poultry flocks.

Environmental Enrichment: Providing environmental enrichment is important for promoting natural behaviors and well-being in free-range poultry. Chickens are naturally curious and social animals that enjoy exploring their surroundings, foraging for food, and engaging in activities such as dust bathing and perching. Providing access to diverse natural features such as trees, shrubs, and vegetation can encourage chickens to exhibit these behaviors and promote physical and psychological health. Additionally, providing enrichment items such as hanging toys, perches, and dust bathing areas can help prevent boredom and reduce stress in free-range poultry flocks.

Chapter 11: Sustainable Chicken-Keeping Practices

Sustainable chicken-keeping practices involve strategies and techniques aimed at minimizing environmental impact, maximizing resource efficiency, and promoting the health and well-being of chickens. By adopting sustainable practices, chicken keepers can reduce their carbon footprint, conserve natural resources, and contribute to a more environmentally friendly and ethical food system. This chapter explores various sustainable chicken-keeping practices, including the use of chicken manure for fertilizer, rotational grazing, integrated pest management, and alternative energy sources.

Using Chicken Manure for Fertilizer

Chicken manure is a valuable source of nutrients and organic matter that can be used to improve soil fertility and enhance plant growth in gardens, orchards, and agricultural fields. Rich in nitrogen, phosphorus, potassium, and other essential nutrients, chicken manure provides plants with a balanced and readily available source of nutrition, promoting healthy growth and abundant yields. However, proper management and application techniques are essential to maximize the benefits of chicken manure while minimizing potential risks and environmental impact.

Benefits of Chicken Manure Fertilizer: Chicken manure offers numerous benefits as a fertilizer, including improved soil structure, increased nutrient availability, and enhanced water retention. When properly composted and applied to soil, chicken manure helps to build soil organic matter, which improves soil structure and promotes aeration, drainage, and root penetration. Additionally, chicken manure releases nutrients slowly over time, reducing the risk of nutrient leaching and runoff, which can contaminate water sources and contribute to pollution. By incorporating chicken manure into soil management practices, gardeners and farmers can improve soil health, increase crop yields, and reduce the need for synthetic fertilizers.

Composting Chicken Manure: Composting is an effective method for converting raw chicken manure into a stable and nutrient-rich fertilizer that is safe to use in gardens and agricultural fields. Composting involves mixing chicken manure with carbon-rich materials such as straw, leaves, or sawdust, along with water and oxygen, to create an aerobic environment where beneficial microorganisms can break down organic matter and convert nutrients into a form that plants can readily absorb. Proper composting techniques, such as turning the compost pile regularly to ensure adequate aeration and moisture, are essential for promoting microbial activity and accelerating the decomposition process. Finished compost should be dark, crumbly, and free of odors, indicating that it is fully decomposed and ready for use as fertilizer.

Applying Chicken Manure Fertilizer: When applying chicken manure fertilizer to soil, it's important to consider factors such as nutrient content, application rates, and timing to ensure optimal results and minimize potential risks.

Chicken manure can be applied as a top dressing or incorporated into soil before planting, depending on the specific needs of the crop and soil conditions. It's essential to avoid applying fresh chicken manure directly to plants, as it can burn roots and leaves due to its high ammonia content. Instead, composted chicken manure or aged manure that has been allowed to decompose for several months should be used to reduce the risk of nutrient imbalances and plant damage.

Managing Nutrient Runoff: One of the key challenges associated with using chicken manure fertilizer is managing nutrient runoff and minimizing the risk of water pollution. Excessive application of chicken manure or improper management practices can result in nutrient leaching, runoff, and contamination of surface and groundwater sources. To mitigate these risks, chicken manure should be applied by soil test recommendations and nutrient management plans, taking into account factors such as soil type, crop needs, and environmental conditions. Additionally, incorporating conservation practices such as cover cropping, crop rotation, and buffer strips can help reduce nutrient runoff and protect water quality.

Integrating Chickens into Permaculture Systems

Integrating chickens into permaculture systems is a holistic approach to sustainable farming that harnesses the natural behaviors and contributions of chickens to enhance soil fertility, pest control, and overall ecosystem health.

Permaculture, derived from the words "permanent" and "agriculture," emphasizes principles such as diversity, resilience, and self-regulation to create productive and resilient ecosystems that mimic natural patterns and processes. Chickens play a vital role in permaculture systems by providing valuable services such as soil aeration, weed control, insect management, and nutrient cycling. This section explores the various ways in which chickens can be integrated into permaculture systems to maximize their benefits and contribute to sustainable agriculture practices.

Benefits of Integrating Chickens into Permaculture Systems

Integrating chickens into permaculture systems offers numerous benefits for soil health, pest management, and ecosystem resilience. Chickens are natural foragers that excel at converting organic matter such as insects, weeds, and kitchen scraps into valuable nutrients for the soil. Through scratching, pecking, and manure deposition, chickens help to aerate the soil, break down organic matter, and distribute nutrients, promoting soil fertility and microbial activity. Additionally, chickens play a crucial role in pest management by consuming insects, grubs, and weed seeds, reducing the need for chemical pesticides and herbicides. By integrating chickens into permaculture systems, farmers and gardeners can create more diverse, productive, and sustainable ecosystems that require fewer external inputs and are more resilient to environmental stresses.

Key Strategies for Integrating Chickens into Permaculture Systems

There are several key strategies for integrating chickens into permaculture systems effectively. One approach is to incorporate chickens into rotational grazing systems, where they are moved regularly between different areas of the farm or garden to provide targeted soil disturbance, weed control, and fertility enhancement. By rotating chickens through pasture areas, orchards, and vegetable beds, farmers can capitalize on their natural foraging instincts and maximize their contributions to soil health and pest management while minimizing damage to crops and vegetation.

Another strategy for integrating chickens into permaculture systems is to design chicken tractors or mobile coops that can be moved easily around the landscape to target specific areas in need of soil improvement or pest control. Chicken tractors are portable enclosures equipped with food, water, and shelter that provide chickens with access to fresh forage while confining them to a designated area. By strategically positioning chicken tractors in orchards, vineyards, or vegetable gardens, farmers can harness the natural behaviors of chickens to improve soil structure, suppress weeds, and control pests without the use of synthetic chemicals.

Complementary Farming Practices

Complementary farming practices are techniques and strategies that work synergistically to enhance the productivity, resilience, and sustainability of agricultural systems. By combining different practices such as crop rotation, cover cropping, and agroforestry, farmers can create more diverse, balanced, and regenerative ecosystems that mimic natural patterns and processes. This section explores various complementary farming practices that complement and enhance the integration of chickens into permaculture systems, including agroecology, polyculture, and regenerative agriculture.

Agroecology: Agroecology is an ecological approach to farming that emphasizes the integration of ecological principles and practices into agricultural systems to promote biodiversity, soil health, and ecosystem resilience. By applying principles such as crop diversity, biological pest control, and soil conservation, agroecological farming systems aim to minimize external inputs such as synthetic fertilizers and pesticides while maximizing the natural processes and services provided by ecosystems. Chickens play a valuable role in agroecological farming systems by contributing to soil fertility, pest management, and nutrient cycling, thereby reducing the need for chemical inputs and promoting a more sustainable and environmentally friendly approach to agriculture.

Polyculture: Polyculture is a farming practice that involves growing multiple crops or species together in the same area to enhance biodiversity, soil fertility, and ecosystem resilience. By diversifying crop plantings and incorporating a variety of complementary species such as legumes, grains, vegetables, and fruit trees, farmers can create more resilient and productive agricultural systems that are less susceptible to pests, diseases, and environmental stresses. Integrating chickens into polyculture systems can further enhance their benefits by providing additional sources of organic matter, nutrient cycling, and pest control, thereby increasing overall system productivity and sustainability.

Regenerative Agriculture: Regenerative agriculture is an approach to farming that seeks to restore and enhance ecosystem health, biodiversity, and soil fertility through a combination of holistic management practices such as cover cropping, crop rotation, and livestock integration. By mimicking natural processes and harnessing the power of photosynthesis to capture carbon and build soil organic matter, regenerative agriculture aims to mitigate climate change, improve water quality, and enhance the resilience of agricultural landscapes. Chickens play a crucial role in regenerative agriculture by contributing to soil fertility, weed control, and pest management, while also providing valuable products such as eggs and meat. By integrating chickens into regenerative farming systems, farmers can harness their natural behaviors and contributions to enhance the overall sustainability and resilience of their operations.

Chapter 12: Legal and Regulatory Considerations

Navigating the legal and regulatory landscape is essential for anyone considering starting a chicken-keeping operation, whether for personal enjoyment or commercial purposes. Understanding and complying with zoning regulations, local ordinances, and other legal requirements is crucial for ensuring the success and legality of your poultry enterprise. This chapter explores the various legal and regulatory considerations that chicken keepers must take into account, including zoning regulations, permitting requirements, and compliance with health and safety standards.

Zoning and Local Ordinances

Zoning regulations and local ordinances govern where and how chickens can be kept within a community or municipality. Zoning laws typically designate specific areas within a jurisdiction for agricultural, residential, commercial, or industrial use, with each zone subject to different regulations and restrictions regarding land use, building codes, and animal husbandry practices. Before starting a chicken-keeping operation, it's essential to research and understand the zoning regulations and local ordinances that apply to your property to ensure compliance and avoid potential legal issues.

Residential Zoning: In many residential areas, zoning regulations restrict or prohibit the keeping of livestock, including chickens, due to concerns about noise, odor, and potential nuisances to neighbors. However, some municipalities have adopted ordinances that allow for the keeping of a limited number of chickens under certain conditions, such as lot size, setback requirements, and coop design standards. These ordinances may also specify regulations regarding the number of chickens allowed per property, coop size, noise mitigation measures, and waste management practices. Before keeping chickens in a residential area, it's essential to review the local zoning regulations and ordinances to ensure compliance and avoid potential conflicts with neighbors and local authorities.

Agricultural Zoning: In agricultural zones, zoning regulations are generally more permissive regarding the keeping of livestock, including chickens, due to the rural nature of these areas and the prevalence of farming activities. However, even in agricultural zones, there may be restrictions regarding setbacks, property boundaries, and coop design to minimize potential impacts on neighboring properties and ensure the health and safety of the chickens. It's important to review the specific zoning regulations and ordinances that apply to agricultural zones in your area to ensure compliance and avoid potential legal issues.

Special Use Permits: In some cases, individuals may need to obtain special use permits or variances to keep chickens in areas where they are not explicitly permitted under existing zoning regulations or local ordinances. Special use permits typically require applicants to demonstrate that their proposed chicken-keeping operation will not create significant adverse impacts on neighboring properties or the surrounding community and may be subject to public hearings, review by planning commissions or zoning boards, and approval by local government authorities. Obtaining a special use permit can be a time-consuming and complex process, requiring careful planning, documentation, and engagement with local officials and stakeholders.

Compliance with Health and Safety Standards

In addition to zoning regulations and local ordinances, chicken keepers must also comply with health and safety standards established by federal, state, and local government agencies to protect public health and welfare. These standards may include regulations regarding the construction and maintenance of chicken coops, sanitation and waste management practices, disease prevention and control measures, and food safety requirements for the production and sale of eggs and meat. Failure to comply with health and safety standards can result in fines, penalties, and legal action by regulatory authorities, as well as potential harm to human and animal health.

Health and Safety Regulations: Health and safety regulations play a critical role in ensuring the well-being of both chickens and humans involved in poultry operations. These regulations are designed to prevent the spread of disease, minimize the risk of foodborne illness, and protect workers from occupational hazards associated with poultry farming. Compliance with health and safety regulations is essential for maintaining the integrity of the food supply chain, safeguarding public health, and promoting responsible stewardship of natural resources. This section explores the various health and safety regulations that apply to poultry operations, including sanitation standards, disease prevention measures, and worker safety requirements.

Sanitation Standards: Sanitation standards are essential for preventing the spread of disease and maintaining hygienic conditions in poultry operations. These standards encompass a range of practices, including cleaning and disinfection of facilities, equipment, and vehicles; proper storage and handling of feed and bedding materials; and management of manure and other waste products. Regular cleaning and disinfection of chicken coops, feeders, and waterers help to control the spread of pathogens such as bacteria, viruses, and parasites that can cause disease in chickens and contaminate eggs and meat intended for human consumption. Proper sanitation practices also help to reduce the risk of foodborne illness and ensure the safety and quality of poultry products.

Disease Prevention Measures: Disease prevention measures are critical for protecting the health and welfare of chickens and preventing the spread of contagious diseases within poultry populations. These measures may include vaccination programs, biosecurity protocols, and monitoring for signs of illness or disease. Vaccination programs help to protect chickens from common infectious diseases such as Newcastle disease, infectious bronchitis, and avian influenza, reducing the risk of outbreaks and minimizing the need for antibiotic treatment. Biosecurity protocols, such as restricting access to poultry facilities, implementing quarantine procedures for new birds, and disinfecting equipment and vehicles, help to prevent the introduction and spread of pathogens between flocks. Monitoring for signs of illness or disease, such as respiratory distress, lethargy, or abnormal behavior, allows poultry producers to detect and respond to health issues promptly, minimizing the impact on flock health and productivity.

Worker Safety Requirements: Worker safety requirements are designed to protect employees from occupational hazards associated with poultry farming, such as exposure to hazardous chemicals, airborne particulates, and physical injuries. These requirements may include providing personal protective equipment (PPE) such as gloves, masks, and goggles to workers, implementing safety training programs, and complying with regulations regarding the safe handling and storage of chemicals and other hazardous materials. Proper ventilation systems and respiratory protection measures help to minimize exposure to dust, ammonia, and other airborne contaminants present in poultry facilities, reducing the risk of respiratory problems and occupational illnesses among workers. Additionally, the ergonomic design of workspaces and equipment can help prevent musculoskeletal injuries and repetitive strain injuries associated with tasks such as lifting, bending, and repetitive motion.

Permits and Licenses

Permits and licenses are required for poultry operations to ensure compliance with regulatory requirements, protect public health and safety, and maintain environmental quality. These permits and licenses may be issued by federal, state, or local government agencies and are typically required for activities such as the construction and operation of poultry facilities, discharge of wastewater and stormwater runoff, and sale of poultry products for human consumption. Obtaining permits and licenses involves submitting applications, undergoing review and approval by regulatory authorities, and complying with applicable regulations and standards. Failure to obtain required permits and licenses can result in fines, penalties, and legal action by regulatory agencies, as well as potential harm to public health and environmental resources.

Construction and Operation Permits: Construction and operation permits are required for the establishment and operation of poultry facilities, including chicken coops, egg-laying facilities, and processing plants. These permits may involve site inspections, environmental impact assessments, and compliance with building codes, zoning regulations, and health and safety standards. Construction permits typically require approval from local building departments or planning commissions and may include conditions such as setback requirements, noise abatement measures, and waste management plans. Operation permits are required to legally operate poultry facilities and may be subject to periodic inspections and compliance monitoring by regulatory agencies.

Environmental Permits: Environmental permits are required for activities that may impact air quality, water quality, or natural resources, such as wastewater discharge, stormwater runoff, and land use changes associated with poultry operations. These permits may be issued by federal, state, or local environmental agencies and typically involve compliance with regulations such as the Clean Water Act, Clean Air Act, and Endangered Species Act. Environmental permits may require the implementation of pollution prevention measures, such as erosion and sediment control practices, nutrient management plans, and wastewater treatment systems, to minimize environmental impacts and protect sensitive habitats and ecosystems.

Food Safety Permits: Food safety permits are required for the production, processing, and sale of poultry products for human consumption to ensure compliance with food safety regulations and standards. These permits may be issued by state or local health departments and typically involve compliance with regulations such as the Food Safety Modernization Act (FSMA) and the Hazard Analysis and Critical Control Points (HACCP) system. Food safety permits may require the implementation of food safety practices such as sanitation and hygiene protocols, temperature control measures, and traceability systems to prevent foodborne illness and ensure the safety and quality of poultry products for consumers.

Chapter 13: Troubleshooting Common Issues

Chicken keeping, like any other agricultural endeavor, comes with its fair share of challenges and obstacles. From health issues to behavioral problems, chicken keepers may encounter a range of common issues that require troubleshooting and problem-solving skills. This chapter aims to address some of the most frequent issues faced by chicken keepers and provide practical solutions to overcome them. By understanding the root causes of these problems and implementing appropriate remedies, chicken keepers can maintain the health and well-being of their flock and ensure a successful and rewarding chicken-keeping experience.

Solving Egg-Laying Problems

Egg-laying problems are a common concern among chicken keepers and can be caused by various factors, including age, breed, nutrition, environment, and health issues. Understanding the reasons behind egg-laying problems is essential for implementing effective solutions and promoting consistent egg production in the flock. This section explores some of the most common egg-laying problems encountered by chicken keepers and offers practical strategies for addressing them.

Age-related Issues: Age-related factors can significantly impact egg production in chickens, with younger hens typically laying more eggs than older ones. As chickens age, their egg production naturally declines, eventually ceasing altogether. Additionally, older hens may experience a decline in egg quality, with eggs becoming smaller, thinner-shelled, and less nutritious over time. While age-related changes are inevitable, providing proper nutrition, healthcare, and environmental enrichment can help support egg production and prolong the productive lifespan of hens.

Nutritional Deficiencies: Nutritional deficiencies can impair egg production and quality in chickens, leading to issues such as decreased egg size, shell abnormalities, and reduced hatchability. Common nutritional deficiencies that may affect egg-laying performance include inadequate levels of protein, calcium, vitamins, and minerals in the diet. Ensuring that chickens are fed a balanced diet that meets their nutritional requirements is essential for promoting optimal egg production and health. Supplementing the diet with calcium-rich foods such as oyster shells or crushed eggshells can help to prevent calcium deficiency and promote strong, healthy eggshells.

Environmental Stressors: Environmental stressors such as extreme temperatures, overcrowding, poor ventilation, and inadequate lighting can negatively impact egg production and quality in chickens. Chickens are sensitive to their environment and may reduce or cease egg production in response to stressful conditions. Providing a comfortable and stress-free environment for chickens is essential for maintaining consistent egg production. This includes ensuring adequate space, ventilation, and lighting in the coop, as well as providing access to clean water, nutritious feed, and comfortable nesting areas.

Health Issues: Health problems such as diseases, parasites, and injuries can disrupt egg production and quality in chickens. Common health issues that may affect egg-laying performance include respiratory infections, reproductive disorders, internal parasites, and injuries to the reproductive tract. Regular health monitoring, preventive healthcare measures, and prompt veterinary attention are essential for preventing and managing health issues in the flock. Addressing health problems promptly can help to minimize their impact on egg production and ensure the overall health and well-being of the flock.

Behavioral Factors: Behavioral factors such as broodiness, aggression, and stress can also affect egg-laying behavior in chickens. Broodiness, the tendency of hens to sit on eggs and incubate them, can disrupt egg production and lead to reduced egg output. Providing suitable nesting areas and discouraging broody behavior through various management techniques can help minimize the impact of broodiness on egg production. Additionally, addressing issues such as aggression within the flock, social hierarchy disputes, and environmental stressors can help to promote a harmonious and conducive environment for egg-laying.

Addressing Aggressive Behavior: Aggressive behavior in chickens can pose challenges for chicken keepers and may manifest in various forms, including pecking, chasing, fighting, and territorial aggression. Understanding the underlying causes of aggression in chickens is essential for effectively addressing and managing this behavior. Aggression in chickens can be triggered by factors such as overcrowding, social hierarchy disputes, environmental stressors, breeding season, and territorial instincts. By identifying the root causes of aggression and implementing appropriate management strategies, chicken keepers can create a harmonious and safe environment for their flock.

Overcrowding is a common cause of aggression in chickens, as it can lead to competition for resources such as food, water, and space. Chickens are naturally social animals, but they also require adequate space and resources to establish and maintain social hierarchies peacefully. Overcrowding can disrupt social dynamics within the flock, leading to increased aggression, bullying, and stress. Providing sufficient space, roosting areas, and environmental enrichment can help to reduce aggression by allowing chickens to establish their territories and social structures without feeling overcrowded or threatened.

Social hierarchy disputes are another common cause of aggression in chickens, particularly among males or mixed-sex flocks. Chickens have a natural pecking order, with dominant individuals asserting their authority over subordinate ones through displays of aggression and dominance. However, when the social hierarchy is disrupted or challenged, it can lead to increased aggression, fighting, and injuries within the flock. Introducing new birds, removing or culling aggressive individuals, and providing multiple feeding and watering stations can help to reduce social hierarchy disputes and minimize aggression among flock members.

Environmental stressors such as inadequate lighting, poor ventilation, extreme temperatures, and predator threats can also contribute to aggressive behavior in chickens. Chickens are sensitive to their environment and may become agitated or defensive in response to perceived threats or discomfort. Providing a comfortable and stress-free environment for chickens is essential for promoting calm and peaceful behavior. This includes ensuring adequate lighting, ventilation, and temperature control in the coop, as well as implementing predator deterrents and security measures to protect the flock from external threats.

Breeding season can trigger aggressive behavior in chickens, particularly among males competing for mates or territory. During the breeding season, hormones such as testosterone can influence behavior and increase aggression levels in males, leading to fighting, aggression towards females, and territorial disputes. Separating aggressive males, providing ample space and resources, and reducing breeding stimuli such as artificial lighting can help mitigate aggression during the breeding season and maintain peace within the flock.

Territorial instincts are natural behaviors exhibited by chickens to defend their territory and resources from perceived threats or intruders. Chickens may become aggressive towards unfamiliar birds, predators, or humans encroaching on their territory, particularly during breeding or brooding periods. Implementing measures to establish and maintain boundaries, such as fencing, netting, and predator-proofing the coop, can help to reduce territorial aggression and protect the flock from harm. Additionally, providing adequate hiding spots, shelters, and escape routes can give chickens a sense of security and reduce stress-induced aggression.

Dealing with Predators

Predators pose a significant threat to the safety and well-being of chickens and can include a wide range of animals such as foxes, raccoons, coyotes, hawks, owls, snakes, and domestic pets. Protecting chickens from predators requires proactive management strategies and the implementation of predator deterrents to minimize the risk of predation. Understanding the behavior and habits of local predators is essential for developing effective predator management plans and safeguarding the flock.

Securing the coop and run is the first line of defense against predators and is essential for protecting chickens from nighttime attacks when they are most vulnerable. The coop should be constructed of sturdy materials such as wood or metal and equipped with secure doors, locks, and latches to prevent predators from gaining access. Windows and vents should be covered with predator-proof mesh or wire to prevent entry, and the coop should be elevated off the ground to deter digging predators such as raccoons and foxes.

Installing predator-proof fencing around the perimeter of the coop and run can help to further deter predators and prevent them from gaining access to the flock. Fencing should be buried several inches underground and extend several feet above ground level to prevent digging, climbing, or jumping predators from breaching the enclosure. Electric fencing or predator-proof netting can also be effective deterrents for aerial predators such as hawks and owls.

Implementing predator deterrents such as motion-activated lights, sound alarms, and predator decoys can help deter predators and alert chicken keepers to potential threats. Motion-activated lights can startle nocturnal predators and make them think twice before approaching the coop, while sound alarms can alert chicken keepers to potential threats and scare off predators. Predator decoys such as fake owls, snakes, or predatory birds can also deter predators by mimicking natural threats and discouraging them from approaching the coop.

Practicing good husbandry and management practices can also help to reduce the risk of predation and protect the flock from harm. This includes securely locking up chickens at night, collecting eggs promptly to prevent attracting predators, and removing potential attractants such as food scraps or garbage from the vicinity of the coop. Regularly inspecting the coop and running for signs of damage or weakness and repairing any vulnerabilities promptly can help to prevent predators from exploiting weaknesses and gaining access to the flock.

Chapter 14: Chicken Products and By-Products

Chicken products and by-products play a significant role in various industries, including agriculture, food production, and manufacturing. From meat and eggs to feathers and manure, chickens provide a wide range of valuable resources that can be utilized in numerous ways. This chapter explores the different products and by-products derived from chickens and their applications in various sectors of the economy.

Utilizing Chicken Manure

Chicken manure is a valuable by-product of poultry farming and is rich in nutrients such as nitrogen, phosphorus, and potassium, making it an excellent source of organic fertilizer for agricultural and horticultural purposes. Chicken manure contains essential nutrients that promote plant growth and soil fertility, making it an ideal choice for organic and sustainable farming practices. However, proper management and handling of chicken manure are essential to maximize its benefits and minimize potential environmental impacts.

One of the primary uses of chicken manure is as a soil amendment and fertilizer in crop production. Chicken manure can be applied directly to fields or incorporated into compost to improve soil structure, enhance nutrient content, and promote plant growth. The high nitrogen content of chicken manure makes it particularly beneficial for crops requiring nitrogen-rich fertilization, such as leafy greens, vegetables, and fruit-bearing plants. Additionally, chicken manure contains beneficial microorganisms that help to break down organic matter and release nutrients, further enhancing its effectiveness as a soil amendment.

In addition to its use in crop production, chicken manure can also be utilized in organic gardening and landscaping applications. Chicken manure can be applied to gardens, flower beds, and lawns to improve soil fertility, promote healthy plant growth, and suppress weed growth. When used as a top dressing or mulch, chicken manure releases nutrients slowly over time, providing a continuous source of nutrition for plants and improving soil moisture retention. Additionally, chicken manure can be used to create compost tea, a liquid fertilizer made by steeping compost in water, which can be applied to plants as a foliar spray or soil drench to promote growth and vigor.

Furthermore, chicken manure can be processed into organic fertilizers and soil amendments for commercial sale to agricultural producers and home gardeners.

Composted chicken manure, pelletized chicken manure, and chicken manure-based fertilizers are popular choices for organic farmers and gardeners seeking sustainable and environmentally friendly alternatives to synthetic fertilizers. These products are typically rich in nutrients, free of harmful chemicals and pathogens, and beneficial for soil health and plant growth. By converting chicken manure into value-added products, poultry farmers can generate additional revenue streams and reduce waste.

In addition to its use as a fertilizer, chicken manure can also be utilized in renewable energy production through anaerobic digestion. Anaerobic digestion is a biological process that breaks down organic matter in the absence of oxygen, producing biogas and nutrient-rich digestate as by-products. Chicken manure can be digested in anaerobic digesters to produce biogas, a renewable energy source consisting primarily of methane and carbon dioxide. Biogas can be used to generate heat and electricity or upgraded to biomethane for injection into natural gas pipelines or use as a transportation fuel. The nutrient-rich digestate produced during anaerobic digestion can be used as a biofertilizer or soil amendment, further enhancing the sustainability and circularity of the poultry farming operation.

However, the management of chicken manure presents certain challenges and considerations, particularly regarding odor control, nutrient management, and environmental protection. Chicken manure contains high levels of nitrogen and phosphorus, which can contribute to nutrient runoff and water pollution if not managed properly. Excessive application of chicken manure to fields can lead to nutrient imbalances, soil acidification, and contamination of surface and groundwater with nitrates and phosphates, posing risks to human health and the environment. To mitigate these risks, poultry farmers must implement best management practices for manure management, including proper storage, handling, and application techniques to minimize nutrient loss and environmental impact.

Furthermore, proper odor control measures are essential to address the potential nuisance associated with chicken manure, particularly in residential areas or areas with high population density. Odor mitigation strategies such as covering manure storage areas, incorporating manure into the soil promptly, and using odor-neutralizing additives can help to reduce odor emissions and minimize complaints from neighbors and community members. Additionally, proper storage and handling of chicken manure are essential to prevent the spread of pathogens and reduce the risk of disease transmission to humans and animals.

Harvesting Feathers and Down

Feathers and down are valuable by-products of poultry farming, prized for their warmth, softness, and insulation properties. Harvesting feathers and down from chickens is a common practice in the poultry industry and can be done through various methods, including manual plucking, mechanical plucking, and molting. Feathers and down are used in various industries, including bedding, apparel, insulation, and upholstery, making them valuable commodities in global markets.

Manual plucking is the traditional method of harvesting feathers from chickens and involves pulling out feathers by hand from the bird's skin. Manual plucking is labor-intensive and time-consuming but can yield high-quality feathers and down with minimal damage. Skilled workers can selectively pluck feathers from different parts of the bird, such as the wings, breast, and back, to maximize yield and quality. However, manual plucking requires precision and care to avoid injuring the bird and ensure a humane and ethical harvesting process.

Mechanical plucking is a more efficient and automated method of harvesting feathers from chickens and involves using specialized equipment such as plucking machines or de-feathering machines. Mechanical plucking machines use rotating rubber fingers or abrasive discs to remove feathers from the bird's skin quickly and efficiently. Mechanical plucking can process large numbers of birds in a short period, making it a preferred method for commercial poultry processing operations. However, mechanical plucking can result in lower-quality feathers and down compared to manual plucking, as the process may damage the feathers or leave behind residue on the skin.

Molting is a natural process in which chickens shed their feathers and grow new ones, typically occurring once or twice a year. Molting can be induced artificially through various methods, including dietary manipulation, photoperiod manipulation, and hormone therapy. During molting, chickens shed their old feathers and grow new ones, resulting in a temporary increase in feathers and down production. Molting can be used to harvest feathers and down from chickens without causing harm or distress to the birds, making it a sustainable and ethical method of feather harvesting.

Feathers and down harvested from chickens are used in a wide range of products and applications, including bedding, pillows, comforters, jackets, sleeping bags, and upholstery. Down, in particular, is highly prized for its lightweight, insulating properties and is often used in premium bedding and outdoor apparel. Feathers are used for stuffing pillows, cushions, and upholstery, providing support, loft, and resilience to the finished product. The quality and characteristics of feathers and down depend on factors such as the bird's breed, age, diet, and living conditions, as well as the harvesting and processing methods used.

The processing and cleaning of feathers and down involve several steps to remove dirt, debris, and oil from the feathers and prepare them for use in various products. Feather processing typically begins with washing and sterilizing the feathers to remove dirt, bacteria, and contaminants. Feathers are then dried, sorted, and graded according to size, quality, and type. Down processing involves separating the down clusters from the feathers, cleaning and sanitizing the down, and fluffing it to restore its loft and resilience. The processed feathers and down are then packaged and shipped to manufacturers for use in bedding, apparel, and other products.

In addition to their use in consumer products, feathers and down are also utilized in industrial applications such as insulation, padding, and filtration. Feathers can be processed into feather meal, a high-protein animal feed ingredient used in poultry, livestock, and aquaculture diets. Feather meal is produced by grinding and drying feathers to remove moisture and make them suitable for animal consumption. Feather meal is rich in protein, amino acids, and minerals, making it a valuable source of nutrition for animals. Feather meal can also be used as a sustainable alternative to synthetic fertilizers in agriculture, providing nutrients and organic matter to improve soil fertility and crop yields.

Processing and Utilizing Chicken Meat

Chicken meat is one of the most widely consumed and versatile protein sources worldwide, valued for its lean texture, mild flavor, and versatility in cooking. Processing and utilizing chicken meat involve various steps, from slaughtering and dressing to cutting, packaging, and preparing for consumption. Chicken meat is used in a wide range of culinary applications, including grilling, roasting, frying, baking, and braising, making it a staple ingredient in cuisines around the world.

The processing of chicken meat begins with slaughtering and dressing the birds, typically done at specialized processing plants or facilities. Chickens are humanely slaughtered using methods such as electrical stunning, mechanical stunning, or controlled atmosphere stunning to render them unconscious before slaughter. After slaughter, chickens are eviscerated, de-feathered, and cleaned to remove internal organs, feathers, and contaminants. The carcasses are then inspected for quality, trimmed, and chilled to reduce microbial growth and maintain freshness.

Once processed, chicken meat is further divided into various cuts and portions to meet consumer demand and culinary preferences. Common cuts of chicken meat include breasts, thighs, wings, drumsticks, and quarters, each offering different textures, flavors, and cooking properties. Chicken meat can be sold fresh frozen, or refrigerated, depending on the intended use and market demand. Fresh chicken meat is typically packaged in vacuum-sealed bags or trays and sold refrigerated, while frozen chicken meat is packaged in plastic bags or cartons and sold in the frozen foods section of supermarkets.

In addition to whole cuts, chicken meat can also be further processed into value-added products such as ground chicken, chicken sausages, chicken nuggets, and chicken patties. Ground chicken is made by grinding chicken meat and can be used as a substitute for ground beef or pork in various recipes. Chicken sausages are made by mixing ground chicken meat with seasonings, spices, and other ingredients and stuffing the mixture into casings. Chicken nuggets and patties are made by forming ground chicken meat into bite-sized pieces or patties, coating them in breading or batter, and frying or baking until cooked through.

Chicken meat is a versatile ingredient that can be used in a wide range of recipes and dishes, from traditional favorites such as fried chicken and chicken soup to international cuisines such as Thai curry and Mexican tacos. Chicken meat can be marinated, grilled, roasted, stir-fried, or stewed, allowing for endless culinary possibilities. Chicken meat is also prized for its nutritional value, being rich in protein, vitamins, and minerals while being relatively low in fat and calories. It is a popular choice for health-conscious consumers seeking lean protein sources as part of a balanced diet.

Processing facilities play a crucial role in ensuring the safety, quality, and integrity of chicken meat throughout the production process. Stringent hygiene and sanitation practices are implemented to prevent cross-contamination and ensure food safety. Chicken meat is subject to rigorous inspection and testing for pathogens such as Salmonella and Campylobacter to minimize the risk of foodborne illness. Additionally, measures such as temperature control, proper handling, and storage are implemented to maintain the freshness and shelf life of chicken meat from processing to consumption.

In addition to its use as a primary protein source, chicken meat by-products such as bones, skin, and offal are also utilized in various ways to minimize waste and maximize value. Chicken bones can be used to make stocks, broths, and soups, providing flavor, nutrients, and body to culinary dishes. Chicken skin can be rendered into fat or used to make crispy snacks and cracklings. Chicken offal such as liver, heart, and gizzards are considered delicacies in many cultures and can be used in soups, stews, and stir-fries.

Furthermore, chicken feathers and down are also utilized in various non-food applications, including insulation, padding, and filling materials. Down, in particular, is prized for its lightweight, insulating properties and is used to fill pillows, comforters, sleeping bags, and outerwear. Chicken feathers are used in upholstery, bedding, and insulation products, providing warmth, comfort, and durability. The processing of feathers and down involves cleaning, sterilizing, and sorting to remove dirt, debris, and contaminants before being used in manufacturing applications.

The processing and utilization of chicken meat involve various steps, from slaughtering and dressing to cutting, packaging, and preparing for consumption. Chicken meat is a versatile and widely consumed protein source valued for its lean texture, mild flavor, and nutritional benefits. Processing facilities play a crucial role in ensuring the safety, quality, and integrity of chicken meat throughout the production process. By-products such as bones, skin, and offal are also utilized in various ways to minimize waste and maximize value. Additionally, feathers and down harvested from chickens are utilized in non-food applications such as bedding, insulation, and upholstery, further demonstrating the versatility and sustainability of poultry farming. With proper processing techniques and responsible utilization practices, chicken meat and its by-products contribute to a diverse range of products and applications, supporting global food security, economic development, and consumer demand for nutritious and sustainable food options.

Conclusion: Reflecting on Your Chicken-Keeping Journey

Embarking on a chicken-keeping journey is not merely about raising a flock of birds; it's about cultivating a deeper connection with nature, embracing a sustainable lifestyle, and experiencing the joys and challenges of caring for living creatures. As you reflect on your chicken-keeping journey, you may find yourself filled with a sense of accomplishment, gratitude, and wonder at the bonds formed with your feathered companions and the lessons learned along the way.

One of the most rewarding aspects of chicken keeping is witnessing the daily rhythms of life in the coop, from the cheerful clucking of hens as they forage for food to the contented purring of roosters as they bask in the sun. Chickens have a remarkable ability to bring joy and laughter into our lives with their quirky behaviors, endearing personalities, and gentle demeanor. Whether they're scratching in the dirt, sunbathing in the dust, or roosting in the trees, chickens have a way of captivating our hearts and reminding us of the simple pleasures of rural living.

Furthermore, chicken keeping offers numerous practical benefits beyond the emotional and psychological rewards. Fresh eggs from your backyard coop are a nutritious and delicious addition to your diet, providing a sustainable source of protein, vitamins, and minerals for you and your family. By raising your chickens, you also have greater control over the quality and provenance of your food, knowing exactly where it comes from and how it's been produced. Additionally, chicken manure can be utilized as a valuable fertilizer for your garden, enriching the soil and promoting healthy plant growth.

However, along with the joys of chicken keeping come certain challenges and responsibilities that require dedication, knowledge, and commitment. From predator attacks and health issues to maintenance and upkeep of the coop, chicken keeping requires ongoing care and attention to ensure the health, safety, and well-being of your flock. Learning to recognize the signs of illness, injury, or distress in your chickens and knowing how to respond effectively is essential for their welfare and longevity. Likewise, implementing predator-proofing measures, providing adequate nutrition and shelter, and maintaining a clean and sanitary environment are all vital aspects of responsible chicken keeping.

Despite the challenges and setbacks you may encounter along the way, each experience offers an opportunity for growth, learning, and adaptation. Whether it's overcoming a bout of illness, dealing with a predator attack, or navigating the complexities of flock dynamics, every challenge you face strengthens your resilience, resourcefulness, and compassion as a chicken keeper. Building a supportive community of fellow chicken enthusiasts, seeking advice and guidance from experienced poultry keepers, and staying informed about best practices and emerging trends in chicken husbandry can all help you navigate the ups and downs of chicken keeping with confidence and competence.

As you look back on your chicken-keeping journey, you may find yourself filled with a profound sense of gratitude for the many blessings and joys that chickens have brought into your life. From the simple pleasures of collecting fresh eggs each morning to the profound sense of connection and belonging that comes from caring for living creatures, chicken keeping offers a rich tapestry of experiences that enriches the soul and nourishes the spirit. Whether you're a seasoned poultry keeper or a novice enthusiast, the bond you share with your flock is a testament to the enduring power of nature to inspire, heal, and transform our lives.

Further Resources and Next Steps

As you continue your chicken-keeping journey, you may find yourself eager to deepen your knowledge, expand your skills, and explore new opportunities for growth and learning. Fortunately, there are numerous resources and avenues available to help you along the way, from books and websites to workshops and community events. Whether you're interested in learning more about breed selection, coop design, health and nutrition, or sustainable practices, there's a wealth of information and expertise waiting to be discovered.

Books and publications are invaluable resources for aspiring and experienced chicken keepers alike, offering comprehensive guides, practical tips, and expert advice on all aspects of poultry husbandry. From beginner's guides and breed profiles to advanced manuals and reference books, there's a wealth of literature available to help you deepen your understanding of chickens and improve your skills as a poultry keeper. Consider exploring titles such as "Storey's Guide to Raising Chickens" by Gail Damerow, "The Chicken Health Handbook" by Gail Damerow, and "The Small-Scale Poultry Flock" by Harvey Ussery for in-depth insights and practical guidance on chicken keeping.

Online resources such as websites, forums, and social media groups are also valuable sources of information, support, and community for chicken keepers. Websites like BackyardChickens.com, The Happy Chicken Coop, and The Poultry Site offer a wealth of articles, forums, and resources on all aspects of chicken keeping, from breed selection and coop design to health care and nutrition. Joining online forums and social media groups allows you to connect with fellow chicken enthusiasts, ask questions, share experiences, and learn from others in the poultry community.

Additionally, attending workshops, seminars, and events hosted by local agricultural extension offices, poultry clubs, and community organizations can provide valuable hands-on experience, networking opportunities, and access to expert advice and resources. Workshops and seminars cover a wide range of topics, including breed selection, incubation and hatching, predator management, and sustainable farming practices. By participating in these events, you can gain practical skills, connect with like-minded individuals, and stay informed about the latest developments in chicken husbandry.

Furthermore, consider joining a local poultry club or association to connect with other chicken keepers in your area, participate in poultry shows and exhibitions, and access resources and support networks. Poultry clubs and associations offer a wealth of benefits, including networking opportunities, educational programs, and social events, as well as discounts on supplies and services. Whether you're a backyard hobbyist or a commercial producer, joining a poultry club can help you connect with others who share your passion for poultry and provide a sense of camaraderie and community.

In conclusion, your chicken-keeping journey is a deeply rewarding and enriching experience that offers countless opportunities for learning, growth, and connection. By reflecting on your experiences, seeking out further resources, and continuing to expand your knowledge and skills, you can enhance your enjoyment and success as a poultry keeper and contribute to the health, well-being, and sustainability of your flock. Whether you're a novice enthusiast or a seasoned veteran, the journey of chicken keeping is a lifelong pursuit filled with wonder, discovery, and endless possibilities. Embrace the journey, cherish the moments, and may your flock continue to bring you joy, inspiration, and fulfillment for years to come.

www.ingramcontent.com/pod-product-compliance
Lightning Source LLC
Chambersburg PA
CBHW050300230526
45471CB00005B/1962